The Green Mirage

The Green Mirage

Why a low-carbon economy may be
further off than we think

John Constable

Civitas: Institute for the Study of Civil Society
London

First Published July 2011

© Civitas 2011
55 Tufton Street
London SW1P 3QL

email: books@civitas.org.uk

ISBN 978-1-906837-30-3

Independence: Civitas: Institute for the Study of Civil Society is a
registered educational charity (No. 1085494) and a company limited
by guarantee (No. 04023541). Civitas is financed from a variety of
private sources to avoid over-reliance on any single or small group
of donors.

All publications are independently refereed. All the Institute's
publications seek to further its objective of promoting the advance-
ment of learning. The views expressed are those of the authors, not
of the Institute.

Typeset by
Kevin Dodd

Printed in Great Britain by
Berforts Group Ltd
Stevenage SG1 2BH

Contents

Figures

Page

Tables

Author

John Constable is the Director of the Renewable Energy Foundation, an independent charity that publishes data and analysis on the energy sector. He was educated in the humanities, reading English at Magdalene College, Cambridge, and subsequently taking his PhD there in 1993. He has since taught at both Kyoto University, Japan, and at Cambridge, where, until 2005, he was a Leverhulme Research Fellow and Senior Research Fellow of Magdalene. In that field he is best known as the leading authority on the twentieth-century philosopher of language and aesthetics, I. A. Richards, whose works 1919-1938 he edited in a ten-volume landmark edition, published in 2001. He is also the co-discoverer, with his colleague the Japanese particle physicist and economist Professor Hideaki Aoyama, of the mathematical distinction between verse and prose in English. Dr Constable has been working in energy policy since 2004 and is known for his data-grounded view that current policy targets for renewables are infeasible, unaffordable, and almost certainly counterproductive.

Acknowledgements

Some years ago Michael Laughton drew my attention to the risk that misconceived energy policies would result in wealth destruction, rather than simple and transitory economic disadvantage, and thus started off the train of thought that has resulted in this book. I am grateful also to Daniel Johnson, the editor of *Standpoint*, for commissioning an article in which some of the ideas presented here were given a first outing in summary form. My colleagues and assistants at the Renewable Energy Foundation have contributed in many ways: Dr Lee Moroney kindly generated data to indicate the forward cost of the current UK renewables subsidies, while our intern research assistants, Ruri Saito, Makoto Hokao, Yutaka Kiyono and Sarah Bell, patiently assembled the library of documents on which the book is based. Special thanks are due to Annabel Ross, whose knowledge of econometrics and the Spanish language was indispensable in clarifying many points in official government documents. The anonymous Civitas reviewers have saved me from both obscurities and errors.

Acronyms

ADP	Accelerated deployment policies
ASIF	Asociación de la Industria Fotovoltaica (Solar Photovoltaic Industry Association)
BAU	Business as usual
BauGB	Baugesetzbuch (German Federal Building Code)
BERR	Department of Business, Enterprise and Regulatory Reform
BMU	Bundesministerium für Umwelt, Naturschutz und Reaktorsicherheit (Federal Ministry for the Environment, Nature Conservation and Nuclear Safety)
CaCC	Campaign against Climate Change
Cmd	Command Paper (Government reports)
CP	Command Paper (Government reports)
CWU	Communication Workers Union
DECC	Department of Energy and Climate Change
DG TREN	Directorate General of Energy and Transport (European Commission)
DWP	Department of Work and Pensions
EEG	Energy Economics Group
ETS	Emissions Trading System (European Union)
FEC	Final Energy Consumption
FTE	Full-time equivalent
G-20	Group of 20 (Argentina, Australia, Brazil, Canada, China, France, Germany, India, Indonesia, Italy, Japan, Mexico, Russia, Saudi Arabia, South Africa, Korea, Turkey, the United Kingdom, United States, European Union)
GDP	Gross Domestic Product
GHG(s)	Greenhouse gas(es)
HC Deb	House of Commons Debate
HGV	Heavy goods vehicle

HPI	Happy Planet Index
IER	Institute for Energy Research
ILO	International Labour Organisation
ILR	Industrial and Labor Relations
IOE	International Organisation of Employers
ISTAS	Instituto Sindical de Trabajo, Ambiente y Salud (United Institute for Employment, Health and the Environment)
ITUC	International Trade Union Confederation
kWh	Kilowatt hour
LEI	Lithuanian Energy Institute
ME	Moderate Exports
MITYC	Ministerio de Industria, Turismo y Comércio (Ministry of Industry, Tourism and Commerce)
MW	Megawatt
MWh	Megawatt hour
NCS	National Climate Service
NEF	New Economics Foundation
NREL	National Renewable Energy Laboratory
OE	Optimistic exports
OECD	Organization for Economic Cooperation and Development
OFC	Overseas Food Corporation
OPEC	Organization of Petroleum Exporting Countries
PCS	Public and Commerical Services Union
PE	Pessimistic exports
PV	Photovoltaic
R&D	Research and development
RASC	Royal Army Service Corps
RES	Renewable Energy Sources

REUK	RenewableUK (formerly BWEA – British Wind Energy Association)
ROC	Renewable Obligation Certificate
RWI	Rheinisch-Westfällisches Institut für Wirtschaftsforschung (Rheinisch-Westfällisches Institute for Economic Research)
SEURECO	Société Européene d'Économie
StrEG	Stromeinspeisegesetz (Electricity Feed-in Law)
TAC	Tanganyika Agricultural Corporation
TSSA	Transport Salaried Staffs Association
TWh	Terawatt hour
UAC	United Africa Company
UCU	University and College Union
UNEP	United Nations Environment Program
WTP	Willingness to pay

Summary

- Little-publicised aspects of economic modelling conducted for the European Union in 2009 and still referred to by the Commission show that the probable impact of renewable energy and climate policies will have only 'slight' net benefits in terms of GDP and employment in 2020, even on the assumption that the EU retains more than 40–50 per cent of the global export market in renewable technologies.

- For some states, such as the United Kingdom, the results are marginal even in this scenario, with a likelihood of relative economic contraction and net employment loss.

- Half of the EU's scenarios report a fall in UK GDP of around 0.05 per cent, with half showing growth of up to 0.1 per cent in 2020, both relative to scenarios in which the renewables policies are abandoned.

- All but one of the scenarios show the UK experiencing net employment losses of over 10,000 to over 30,000 as compared to the situation where the renewables policies are cancelled.

- The Commission describes EU-wide employment benefits in terms of gross effects of around three million jobs by 2020, but the study's fine detail reveals that the net employment effects, even over the whole EU-27, are numbered in the low hundreds of thousands, as the higher costs of energy destroy jobs in other sections of the economy.

- Overall EU-27 GDP growth as a result of the renewables policy is predicted, even in the unrealistically optimistic export scenarios, to be less than 0.5 per cent in 2020, a figure that is marginal and well within the measuring error.

- At best, these are small net benefits for what the gross employment figures indicate is a major and inorganic rebalancing of the European economy fraught with significant social, technological, and economic risks.

- Examination of empirical evidence from Germany and Spain confirms these risks, and strongly suggests that the net economic effects of subsidising renewables to meet targets will suppress activity in other non-energy sectors of the economy, effects that will predominate over time, particularly if the EU does not maintain a high share of the world export market in renewables.

- The experience of competition between the German and Chinese solar industries does not encourage the view that Europe can dominate the renewables market.

- Current policies and much political pressure supporting the green economy agenda in the UK and elsewhere are premised on high levels of governmental coercion and state management of the energy sector, often invoking analogies with wartime production where cost is no object – a fact that arguably invalidates the comparison.

- Concern about the desirability of economic planning of this kind and the inherent wisdom of the measures is obscured since much support for contemporary proposals for the low-carbon economy comes from extremist perspectives that either welcome high levels of state involvement in order to prosecute a programme of social justice or embrace economic contraction on the questionable grounds that it will enhance public well-being.

- Mainstream political representatives both in the UK and the EU tend to underplay the degree of state management required, and minimise the risks of relative economic contraction, while endorsing policy measures that differ little from those recommended by fringe thinkers.

- Significantly, mainstream politicians and extremists alike draw explicit inspiration from the New Deal measures of President Franklin Delano Roosevelt in the 1930s. However, such a comparison is not encouraging, as there is a growing body of analytic historical economics suggesting that the net effect of the New Deal policies was economically negative and succeeded in prolonging the Depression.

- Examination of one of the UK government's principal attempts to manage a large business enterprise, the Groundnut Scheme to grow peanuts in East Africa in the late 1940s, reveals interesting parallels with current proposals for a low-carbon economy, including an inappropriate conception of the project as being quasi-military in character (and thus cost indifferent), a mistaken belief that there was no technology risk and an unwillingness to assimilate evidence of incipient failure in time to take corrective action.

- In conclusion, the European Union's target-led, state-managed and subsidy-driven policies are likely to cause the premature adoption of sub-optimal and costly technologies, exhibiting low productivity, with resulting net economic contraction, relative to the alternatives, and wealth destruction.

- In the period April 2002 to March 2010 the UK spent £5.6 billion subsidising dedicated renewable electricity plant, at a cost of £200,000 per wind industry worker. Subsidy per wind industry worker in the year April 2009 to March 2010 amounted to £57,000, which is greatly in excess of the median earnings in either the public (£29,000) or the private sectors (£25,000). While it is not yet possible to estimate the net employment impacts of such costs, they seem unlikely to be positive.

- Revisions to policy are essential if future invention and innovation are to be encouraged and thus stand some chance of delivering technologies that might substantiate a fundamentally economic and organic transition to low-carbon prosperity.

- If government projections for renewable electricity growth are fulfilled to 2020, a further £39 billion of subsidy cost due to the Renewables Obligation will be incurred and added to consumer bills in the period 2011 to 2020.

- Even if we assume that no further growth is required by policy after 2020, the cost of continued support to generators built in the previous decade would add a further £60 billion, giving a total cost for the Renewables Obligation in the period 2002 to 2030 of around £100 billion.

Introduction

It is a rare policy in the economic strategy of any OECD state that does not come with promises of parallel contributions towards meeting environmental goals. But the public remains unsure both of the sincerity and the substance of these offers, and responds with a wide range of sceptical enquiries, ranging from the technical and the philosophical to the reductively flippant. *Will environmentalism enhance employment opportunities through low-carbon growth? Is the green economy feasible, and its achievement really a matter of political will-power only? Is CO_2-free growth a speechmaker's fantasy, forever out of reach like the crock of gold at the foot of the rainbow? Are green times just around the corner? Low-carbon jobs: myth or reality?* There seems little doubt that such questions, even when presented with apparent cynicism, are often asked in the hope that the answers are straightforwardly positive. It is easy enough to give a trivially comforting response. A green future is *conceivable*, simply through an act of imagination that supposes the global economy at some future date to function without the emission of greenhouse gases. By definition all the jobs in such an economy will be green. We need make no assumptions about the technologies involved, just suppose that they exist and have been adopted, whatever they are; waste-free nuclear generation on the broadest scale, perhaps. Such a vision may be speculative, but it is not incompatible with our understanding of the physical world, the laws of thermodynamics for example. But a response of this kind does no more than recapitulate the hazy and rhetorical propositions that provoked the anxious interrogations in the first place. To be told that there is no fundamental or absolute obstacle to an economy driven by energy and industrial processes that emit no greenhouse gases is unsurprising. We never doubted it; that wasn't what we were asking about. Our deepest doubts are not ontological but practical. We want to know about the difficulties along the way, and what sort of society lies at the other end of the transition. On this point, notoriously, governments are unavoidably vague.

It is rather as if we are standing on one mountain range, looking across at the foothills of another, the peaks of which are wrapped in cloud. If asked to climb those peaks, we might respond by wondering

xvii

not only about the mysterious peak itself, but the nature of the land between, and how this territory is to be traversed. Granted that the low-carbon transition is not science fiction and can it be made in the timescale projected, what will it be like when we arrive? The first of those questions has scientific aspects, involving the nature of the technology and engineering required for the transition, not to mention economic concerns relating to cost– in other words the sacrifices we would have to make *en route*. These are important perspectives, but even this framing of the issue tends to obscure the deepest source of unease, namely the character of the as yet obscure low-carbon economy.

Simply, it is rational to wonder what demographic and socio-political characteristics would be exhibited by a green society attainable with current policies. For example, how many people would be supported in and during the transition to such an economy, at what living standards, and with what levels of individual and political freedom? Any discussion of the green-collar future that leaves these questions unaddressed will be necessarily frustrating, however rich the consideration of potential routes or difficulties, and however rigorous the technical economic analysis.

To put it simply, the haze of questions raised by the public about the green future can be summarised as doubts as to whether the rising trends in contemporary levels of wealth and progressively more liberal political settlements around the world can be sustained without the emission of greenhouse gases and the consumption of finite energy resources. We want to know if 'business as usual' can be prolonged, or whether deep changes in economic and socio-political organisation are a pre-requisite to the reduction of environmental impact.

On the whole, mainstream politicians wish to intimate, without actually asserting, that their environmental ambitions are not only compatible with our desire for wealth and freedom, but perhaps an improvement on the current dispensation. A minority, usually on the green left, suggests that this is untrue, and that radical transformation entailing abandonment of liberal individualism, industrial consumerism, and a drastic reduction in levels of personal wealth that is not only unavoidable, but is so desirable in itself that it constitutes an overwhelming argument in favour of the low-carbon transition quite aside from environmental considerations.

The current analysis will conclude that while conventional politicians are correct in suggesting that there is no necessary incompatibility between environmentalism and what we might for convenience call the Western ideal, they are wrong in suggesting that this idea is unthreatened by *current* clean energy technologies and the transition path mapped out by contemporary policies. This latter point is understood by the green left, which correctly argues that only governmental coercion entailing reductions in personal freedom and wealth can deliver a low-carbon economy founded on currently available technologies. However, these greens are unrealistic in assuming that this is politically sustainable; indeed, there seems to be a degree of insincere and attention-seeking perversity in endorsing positions that so obviously fly in the face of all but universal human desires.

Furthermore, both these polar positions are quite mistaken in thinking that a clear, safe and certain route map exists to move us from where we are today to where we might be in the greener future. Even cursory familiarity with the state of low-carbon industry and energy generation reveals that the level of science and engineering is primitive, and that the technology risk of current policies is high, with the clear danger that as current endeavours fail they will produce counterproductive outcomes, involving greater not lesser levels of environmental degradation.

Put another way, the environmental and resource erosion problems that confront us are too complex and poorly understood to permit us to do more than anticipate their character and our likely response for more than a few years into the future. With the possible exception of nuclear fission, about which there is much disagreement because of its residual effects, contemporary clean or green technologies cannot deliver a continuation of the current social and economic ascent in an environmentally sustainable fashion, and long-term solutions, if there are to be any, must come from incremental invention and subsequent innovation. The only honest answer to the agonised question as to whether we can build a high road to the bright green future is that we have only the vaguest idea of the general direction of travel, and consequently there is little or no possibility of being specific in our plans of how to get there or what materials will be needed to complete the journey in safety and with our valuables intact.

Consequently, much of the discussion in the following study is intended to undermine confidence in the projections of consensual politics and indicate that the resulting econo-environmental policies are predicated on a degree of long-term planning that is quite inappropriate given the radical uncertainties involved. Imagine a forward-thinking monarch in 1700 attempting to plan the conduct of his own state to ensure development towards a network of global societies and economies that can sustain 6,000 million people three hundred years in the future. Such a person would be best advised to pay no attention to speculation on distal concerns, but instead to concentrate on the solution of proximal difficulties with the aim of exploring the conceptual space near at hand. By looking too far ahead such a planner would have tripped over obstacles that could have been revealed by less ambitious progress, and would, almost certainly, have overlooked openings that would have proved to be more rewarding than anything that might have been anticipated. Our own situation is no different, and there are grounds for guarded optimism. Dead ends cannot be ruled out, and history records many, but in the last three hundred years epistemologically and economically free societies have produced advances in understanding and the application of knowledge to which most of us owe our existence. There is, then, reason to think that if those forces are permitted to work on the environmental question, incremental solutions, perhaps inconceivable at present since we misunderstand the problem, will progressively be found. By contrast contemporary efforts to plan a low-carbon economy with full or high employment will not result in a tolerable transition or a satisfactory outcome, and may simply fail.

PART ONE

The Politics of the Low-carbon Economy

Chapter One

The 'Triple Win'

Introduction

Discussions of the potential for a green or low-carbon economy vary greatly in character and depth. There is no single vision, and features of one presentation will be inconsistent with those of another. This is partly because political circumstances vary from country to country, but also due to differences of perspective that are as distinct in fundamental or long-term aims as Baptist and Bootlegger, though they may agree on short-term policy.

However, reviewing major publications, as well as many minor and echoic effects in the press both general and industry-specific, it is possible to see one general feature common to nearly all of them, namely the *multiple win*, which is the view that policies for a green economy succeed in reconciling or simultaneously achieving a wide range of goals, some of which might be thought to be incompatible and at least difficult to achieve in tandem.

A representative example might be taken from the International Labour Organisation (ILO), itself a significant source of policy theory and comment in this area. Announcing a seminar in Copenhagen in December 2009, the ILO entitled its press release: 'Economic recovery and Green Jobs: win-win for development, climate and labour?'

> During the last few months many countries have approved the so-called stimulus packages, with a view to lead these countries towards a quick economic recovery. Many of these countries have taken this opportunity to shift towards a more sustainable development path, by reducing GHG emissions, producing renewable energies and other environmentally-friendly measures while creating Green Jobs and tackling social inequalities.[1]

Such a 'Win-Win' is inherently attractive, though those familiar with the difficulties in simultaneously maximising two variables may feel a little less enthusiastic and regard this phrasing more as an elegant trope than a discovery in economic theory. In the real world a trade-off is inevitable, but this is an important and common type of phrase, encapsulating the hopes of many politicians. British Prime

Minister David Cameron, here commenting on offshore wind in a speech to the Confederation of British Industry, is not unusual:

> It's a triple win. It will help secure our energy supplies, protect our planet and the Carbon Trust says it could create 70,000 jobs.[2]

A more comprehensive statement, and perhaps the inspiration for Mr Cameron's speechmakers, is to be found in this remark by the then EU Commission Vice-President Günter Verheugen, in a statement made on 30 November 2007, which builds on earlier statements of intent from Stavros Dimas, European Commissioner for the Environment, committing the EU to a 'new industrial revolution':[3]

> Tackling climate change requires a varied, integrated approach, which addresses the triple goal of competitiveness, energy and the environment. European leaders have made clear that Europe intends to lead the move to a global low-carbon economy. The opportunity is to be the technology leader and therefore technology supplier in the future. Taking action now can give European businesses an advantage over others, so they manufacture the safest and cleanest products, which the world is waiting for.[4]

Nor is this approach confined to Europe. In an article pleasingly entitled 'Millions of Jobs of a Different Collar', a correspondent for the *New York Times* noted that:

> Presidential candidates talk about the promise of 'green collar' jobs – an economy with millions of workers installing solar panels, weatherizing homes, brewing biofuels, building hybrid cars and erecting giant wind turbines. Labour unions view these jobs as replacements for positions lost to overseas manufacturing and outsourcing. Urban groups view training in green jobs as a route out of poverty. And environmentalists say they are crucial to combating climate change.[5]

Such remarks have appeared frequently in speeches by President Obama, but we need not seek far for further instances of this thesis, for it is truly a global commonplace to assert that governmental action in order to prosecute environmental ends, specifically climate change mitigation policies, will have numerous other benefits, including economic regeneration and the promotion of a progressive social agenda. For example, Greg Barker MP, Minister of State for Climate Change, can be found observing that:

> Crucially, the Green Deal will be great for jobs. Were all 26 million households to take up the Green Deal over the next 20 years, employment in the sector would rise from its current level of 27,000 to something approaching 250,000, working all

around the country to make our housing stock fit for a low-carbon world. Insulation installers and others in the retrofit supply chain all stand to benefit from this long overdue energy efficiency makeover.[6]

Furthermore, though this is a subtle feature of its rhetoric, there is an implication that not only are there multiple simultaneous wins or winners, but that no one loses and there are no losses. The practical utility of such a vision for the middle-ground consensus-building politician should be sufficiently obvious, though even brief reflection reveals that while it has an initially intuitive plausibility, it can reverse, necker cube fashion, and appear quite unconvincing.

No Losers?

As a means of calibrating our understanding of the mainstream political position, it is worth pausing to review an extremist thesis that is compelled by its own political circumstances to confront the inherent oddities of the 'no losers' assumption. The Campaign against Climate Change (CaCC) is a London-based environmental campaign with a broad left-oriented political support base. George Monbiot is Honorary President, and Michael Meacher MP and Caroline Lucas MP are Vice Presidents.[7] A principal goal of the organisation is to develop broad-based union movement support for the environmental agenda. As part of this attempt the CaCC Trade Union group has published a pamphlet *One million climate jobs: solutions to the economic and environmental crises* (2010), with a set of supporting technical papers available online.[8] The authors of this set of documents, who include Barbara Harris-White, Professor of Development Studies at the University of Oxford, explicitly recognise that the low-carbon transition threatens employment in existing industries, the pamphlet granting that 'Important groups of workers now fear for their jobs in the new economy'.[9] One of the supporting papers, 'Jobs Gained and Lost', calculates that an ambitious transition to renewable energy and low-carbon transport will cause the loss of 594,000 jobs over twenty years, 326,000 of these being in the motor vehicle industry, 90,000 in air transport, 180,000 in road freight and 98,000 in energy.[10] It is worth noting that these numbers are direct losses only, no account being taken of the effect of policy-induced cost increases on the inputs to other economic activities. Nevertheless, even as it stands, this estimate is candid to the point of indiscretion. The authors recognise that it will be hard to gain support for policies that have such significant and negative impacts on sectors rich in union members, and the fundamental drive of the pamphlet is to argue that any losses arising from the low-carbon transition it proposes can be more than recouped. However, the authors can think of no other solution than a state-instituted and deficit-funded 'National Climate Service':

> No one will lose out. Of course some people are going to lose their jobs in a low-carbon economy. But a National Climate Service can have a simple policy. Anyone who loses their job because of the new economy will be offered work in the NCS, with retraining and their old wages guaranteed.[11]

This National Climate Service will, the pamphlet tells us, 'do the work that needs to be done' in reducing the United Kingdom's carbon footprint, from operating renewable energy generators, to refitting buildings and putting 'buses on the streets':

> Luckily, many of the jobs available in the National Climate Service will be ones that fit the skills, experience and lifestyles of the people who are losing the old jobs. Aircraft manufacture and engine manufacture are not that different from turbine manufacture. Displaced HGV drivers will be able to retrain to drive buses, vans, long-distance coaches and trains. They will also be able to work on sea-going vessels, in ways that demand endurance, concentration, and long distance travel. The people displaced from the oil and gas industry have many of the skills needed for offshore energy work.[12]

There would be about a million such employees at any one time, which is roughly the number of full-time-equivalent jobs in the NHS.[13] CaCC estimates that this would cost £52bn a year, including materials and NI and pension contributions.[14] However, these figures are questionable. In the financial year 2009/2010 the NHS cost in the region of £99.8bn,[15] of which approximately two-thirds would have been spent on salaries and staff costs.[16] With this in mind, and given the capital-intensive nature of renewable energy, it is clear that CaCC's figures are probably an underestimate. The proposals seem unlikely to gain much genuine traction, even with the more extreme of the major unions, but it would be a mistake to dismiss these views as insignificant. CaCC's infeasible and dangerous suggestions are clumsy but recognisable cousins, at some removes, of the rhetoric of Mr Cameron, Mr Barker and Mr Verheugen.

Decent Work

Fortunately, there are more definite and pedestrian elaborations of these assumptions that crystallise previous arguments and inspire subsequent developments. Perhaps the most important of these is work by the Worldwatch Institute and the ILR School of the Global Labour Institute at Cornell University for the United Nations Environment Program (UNEP), which was published as *Green Jobs: Towards decent work in a sustainable, low-carbon world*, a study also supported by the International Labour Organisation (ILO), the International Organisation of Employers (IOE) and the International Trade Union Confederation (ITUC).

This study is extensive and can be said to function as a clearing-house, bringing earlier scattered and casual references to the economic benefits of renewable energy and other low-carbon technologies into contact with a strong tradition of left-leaning international development theory and politics. As the authors remark at the outset: 'Many declaim a future of green jobs – but few present specifics',[17] and by virtue of this greater degree of engagement the UNEP work is obliged to touch on what will be an enduring theme in the current study: net impact. UNEP observes that 'employment will be affected in at least four ways as the economy is oriented toward greater sustainability':

> First, in some cases, additional jobs will be created – as in the manufacturing of pollution-control devices added to existing production equipment.
>
> Second, some employment will be substituted – as in shifting from fossil fuels to renewables, or from truck manufacturing to rail car manufacturing, or from landfilling and waste incineration to recycling.
>
> Third, certain jobs may be eliminated without direct replacement – as when packaging materials are discouraged or banned and their production is discontinued.
>
> Fourth, it would appear that many existing jobs (especially such as plumbers, electricians, metal workers, and construction workers) will simply be transformed and redefined as day-to-day skill sets, work methods, and profiles are greened.[18]

Notably absent from this list of impacts is the potential suppression of economic activity and thus employment in other sectors of the economy affected by the higher costs implied when inputs are provided from sources favoured on environmental grounds.

Admittedly, item three seems to sound a muted warning note, but the authors offer the consolation that although 'some workers may be hurt in the economic restructuring… winners are likely to far outnumber losers'. However, no attempt is made to quantify these points and the authors offer no concrete recommendations for dealing with emergent problems, the responsibility for their solution being perfunctorily transferred to a non-specific system of socialised benefit: 'Public policy can and should seek to minimize disparities among putative winners and losers… and avoid these distinctions becoming permanent features.' While the problem might be soluble in this ad hoc way if it was of only small scale, and the resulting economy were prosperous, it is not unreasonable to wonder if it would prove to be as simple a matter if the losers were numerous and the supporting economy only generating modest levels of wealth.

Nevertheless, UNEP's definitions and orienting positions have the great virtue of making explicit what is silently and often, doubtless, unknowingly assumed. Take, for example, the following key paragraph:

> A successful strategy to green the economy involves environmental and social full-cost pricing of energy and materials inputs, in order to discourage unsustainable patterns of production and consumption. In general, such a strategy is diametrically opposite to one where companies compete on price, not quality; externalize social and environmental costs; and seek out the cheapest inputs of materials and labor. A green economy is an economy that values nature and people and creates decent, well-paying jobs.[19]

The concluding sentence is an elegant and clear statement of position that would find few dissenters amongst politicians toying with environmentalist positions, but it nevertheless embeds striking assumptions. There is no clear reason for thinking that a green economy would necessarily create decent, well-paying jobs; indeed, it is not difficult to see that valuation of 'nature' might require that 'people' take second place, and some thoroughgoing, deep green environmentalists would so argue. Even if we accept that we would prefer a green economy to involve pleasant working conditions and good wages, it is not at all clear how this is to be combined with a policy that increases the costs of inputs such as energy without either increasing the final cost to consumers, thus suppressing consumption and further economic activity, or exerting a downward pressure on wages.

Indeed, the UNEP study at many points reveals a tendency, common in many similar studies, to bind together in a forced unity qualities that enjoy no necessary relation, as in the following paragraph:

> Green jobs need to be decent work, i.e. good jobs which offer adequate wages, safe working conditions, job security, reasonable career prospects, and worker rights. People's livelihoods and sense of dignity are bound up tightly with their jobs. A job that is exploitative, harmful, fails to pay a living wage, and thus condemns workers to a life of poverty, can hardly be hailed as green.[20]

However distasteful it might seem, there are simply no grounds for thinking that a green job cannot be poorly paid, dangerous, insecure, and exploitative. Indeed, UNEP knows this, since they concede in the next paragraph:

> There are today millions of jobs in sectors that are nominally in support of environmental goals – such as the electronics recycling industry in Asia, or biofuel feedstock plantations in Latin America, for instance – but whose day-to-day reality is characterized by extremely poor practices, exposing workers to hazardous substances or denying them the freedom of association.[21]

The only basic necessary and sufficient condition of a green job, a notoriously vague concept in the literature in any case, is that it does not harm the environment. That condition can be fulfilled quite independently of other qualities, including wages, iniquitous contractual arrangements, restriction of freedom of association, and even exposure to dangerous substances, if that exposure is a necessary means to prevent the release of those substances to the wider environment. That is to say, all other considerations apart from this basic condition can quite consistently be regarded as supplementary matters. They may be desirable in themselves, but they are not integral to the nature of the green economy.

Jobs for All

A reluctance rigorously to separate such independent matters pervades much comment on the low-carbon or environmentalised future, and it is a prominent hallmark of the UNEP study. The most salient of these is the desire to provide jobs, indeed jobs for all. The study argues:

> Economic systems that are able to churn out huge volumes of products but require less and less labor to do so pose the dual challenge of environmental impact and unemployment. In the future, not only do jobs need to be more green, their very essence may need to be redefined.[22]

The confusion here is deep. Economic systems that provide goods and services with less labour are making wealth available to progressively larger sections of the population because the production cost of those goods and services falls as productivity rises. But a failure to grasp this point leads the UNEP study, and many others writing in a similar vein, to cite one of the major failings of renewable energy as if it were a risk-free bonus:

> Along with expanding investment flows and growing production capacities, employment in renewable energy is growing at a rapid pace, and this growth seems likely to accelerate in the years ahead. Compared to fossil-fuel power plants, renewable energy generates more jobs per unit of installed capacity, per unit of power generated and per dollar invested.[23]

With evident excitement the study remarks on the possibility that some 2.1 million people might be employed in wind energy, 6.3 million in solar PV, and 12 million in biofuels by 2030.[24] However, if there is less energy produced per employee from such systems, compared to alternative fuels, and high wages are provided, as UNEP insists is a condition of a true green job, then it is inevitable that the cost of that energy will be relatively high, reducing activity in the rest of the economy. It is conceivable, then, that the number of losers will not be small, as the UNEP study had initially hoped, but could be considerable.

This failure to connect higher levels of employment in the green sector with higher costs, and thus with suppressed economic activity elsewhere, is evident in other parts of the text, for example when the study authors note that:

> The increase in demand for green building components and energy-efficient equipment will stimulate green manufacturing jobs. Energy-efficient equipment often requires more skilled labor than their inefficient counterparts, thus leading to not only a larger number of jobs, but also higher-skilled, higher-paying employment.[25]

But if such energy-efficient equipment is more expensive, and fuel is cheap, then its use may not be economically or environmentally sound (since a higher cost is roughly isometric with resource use). Such subtleties seem to escape the UNEP authors, but reasoning of this kind is not unusual in the green economy literature. Similarly, it is interesting to note the study's apparent bafflement when describing developments in transport:

> Railways are more environment-friendly and labor intensive than the car industry. But the trend over the last few decades has been away from railways in many countries, and employment – both in running rail lines and in manufacturing locomotives and rolling stock – has fallen accordingly. Even in China (where the rail network grew by 24 percent in 1992–2002) railway employment was cut from 3.4 million to 1.8 million. India's railway jobs declined from 1.7 million to 1.5 million. In Europe, railway employment is down to about 900,000 jobs; the number of workers in manufacturing rail and tram locomotives and rolling stock there has declined to 140,000. A sustainable transport policy needs to reverse this trend. A strategic investment policy to build and rebuild rail networks, integrating high-speed inter-city lines with regional and local lines would offer a substantial expansion in green jobs.[26]

However, we may wonder whether a process that is more labour intensive is in its full extent more environmentally friendly, since its cost includes the support and environmental impact of its employees. Putting this aside, for the sake of simplicity, it is in any case remarkable that UNEP finds it surprising that even railways seem to be endeavouring to employ fewer people, in other words to improve their efficiency. There might be safety concerns that would motivate us to argue that this trend should be reversed, but there can be no such economic argument, and if there is an argument on the grounds of sustainability it would be honest to explain the consequences, namely either higher ticket prices or lower wages.

Against the Market

The pattern that emerges from this sequence of quotations is one in which the creation of jobs is seen as an end in itself, and this end is amalgamated with the environmental character of the project, as if they were bound together by some mutual entailment. However, a business operation aims to attract customers by providing a good or a service in as cost-effective a form as it can; higher levels of employment for a given quantity of product or level of quality are a mark of inefficiency, higher cost, and thus relative unaffordability.

Fortunately, there is no reason for believing that environmentally promising technologies or processes and the businesses based on them are *necessarily* economically inefficient in comparison with the alternatives. Current incarnations may be so at present, employing more people and consuming more resources, but those processes may in time become more efficient. Conversely, conventional processes may become more expensive: for example, fossil fuels may become harder to extract. Such changes lie in the uncertain future, and for the foreseeable future it seems that if cleaner technologies are to be adopted, then we must accept that they will either pay their employees less than current alternatives, or that they will cost rather more, with all the significant economic impacts that this implies, certainly in the energy sector.

The UNEP study, however, is, like many advocates of a green economy, frustratingly confused on this point. Having admitted that environmentally friendly energy production and energy efficient devices require more labour than their competitors, UNEP inconsequentially remarks that: 'Green innovation helps businesses stay at the cutting edge and hold down costs by reducing wasteful practices. This is essential for retaining existing jobs and creating new ones.'[27] In fact, UNEP is quite aware that such normal economic considerations will not drive a green jobs revolution at present, and they immediately note the need for government coercion:

> However, the risk and profit appraisals typical of business, the seemingly ever-rising expectations of shareholders, as well as concerns about protecting intellectual property, may together impede the flow of capital into the green economy. On the basis of current experience in various areas – from vehicle fuel economy to carbon trading – it appears that a purely market-driven process will not be able to deliver the changes needed at the scale and speed demanded by the climate crisis. Forward-thinking public policies remain indispensable in facilitating and guiding the process of greening business. Governments at the global, national and local levels must establish an ambitious and clear policy framework to support and reward sustainable economic activity and be prepared to confront those whose business practices continue to pose a serious threat to a sustainable future. Timely action on the scale needed will occur only with a clear set of targets and mandates, business incentives, public investment, carbon or other ecological taxes, subsidy reform, sharing of green technologies, and scaling up and replicating best practices through genuine public-private partnerships. With progress on these fronts, millions of new green jobs can indeed be generated in coming years.[28]

This paragraph forms the principal conclusion of the UNEP study, and may be said to encapsulate in convenient form the subtext of a large part of the green jobs literature, which is similarly a confusion of apparently market-oriented language – 'flow of capital', 'business incentive' – and a willingness to contemplate universal and unrestricted economic planning, from the local to international levels, in order to deliver favoured outcomes.

Significantly, this literature exhibits a widespread indifference to doubts around innovation and technology risk; it is assumed that all the necessary technologies now exist, will work as intended, and will improve steadily and spontaneously. None of those assumptions is correct, and planned economies have a poor track record of developing and applying novel technical solutions to complex problems. Even if we waive this as mere detail, which it is not, there is a troubling lack of concern at the political and social dangers of a close alliance between the state and the producers that are to be its delivery vehicles; the interest of the consumer is rarely if ever mentioned; the likelihood of gross inefficiency leading to failure is undiscussed; the probability of corruption is never contemplated.

While it would be misleading to assume that every micro-proposition in the UNEP study also applies to, for example, Mr Cameron's programme for green growth, or that of the European Union, there are sufficient general or macroscopic similarities to justify the proposal that the Worldwatch analysis is clumsily transparent where governments use ambiguous gesture to finesse a situation that would become embarrassing if drawn out in specifics. On the issue of multiple wins, UNEP is entirely consistent with the positions already noted as characteristic of much consensus political argument for the green economy. It is, indeed, a key plank of green economy theory as it currently stands that green development will produce economic growth, and that the historical trend towards greater overall wealth will be maintained within the low-carbon economy. As we have already noted by examining various positions in UNEP's study, this view is not self-evidently correct or even internally consistent, but it is overwhelmingly the majority position, and even informs highly eccentric statements such as that of the Campaign against Climate Change. For the most part these views are probably sincere, though some politicians may well be judging that the public is not yet ready to be asked to accept a decline in living standards for the sake of the environment, as they might, for example, in a time of war when faced with a manifest external threat.

Chapter Two

Embracing Wealth Destruction

While mainstream institutions prefer tactful evasions, there are advocates for a green transition that make little or no attempt to conceal their belief that a green economy will of necessity bring about a radical alteration in the current economic trajectory, and that far from needing an apology this is to be embraced as a positive development. While this is a minority and even an extremist position, it is not only internally consistent but is in some ways more realistic than the general consensus, and it deserves examination if only to bring out still more clearly the character of the received wisdom implied in positions offering a 'triple win'.

A War Footing

A thoroughly worked out and lucid example can be found in the writings of Andrew Simms of the New Economics Foundation (NEF), particularly in *The New Home Front* (2011), and in *A Green New Deal* (2008).[1] The most recent of these was written for Caroline Lucas, the leader of the UK's Green Party and its sole MP, and forms the launch manifesto of the New Home Front initiative, which aims to engage with those who remember the austerity measures of the 1939–45 war, and draw on this as an inspiration for a new climate change policy. Its grounding hypothesis is that, to use the words of the earlier study, there is a 'need for mobilisation as for war':[2]

> If we are to overcome the threat of climate change, our country will need to move onto the equivalent of a war footing, where the efforts of individuals, organisations and government are harnessed together and directed to a common goal. Only this will provide the urgency, energy and creativity we need to avert disaster.[3]

It is interesting to reflect that expenditure in wartime is not usually constrained by normal budgetary considerations; the need for victory is beyond doubt and acute, and the problem faced by the government is simple, usually a hostile nation state. On the other hand, climate change, environmental pollution and resource erosion are complex and chronic problems, the characterisation of which is controversial in ways that have significant implications for the manner in which we address them: for example, the balance between adaptation and mitigation. War is a flawed analogy, and misleading, since it short-circuits discussions about costs and benefits that are crucial to a successful approach to our difficulties in energy and environment.

Nevertheless, there is a growing body of environmentalist literature that employs this conceptual vehicle, *One million climate jobs* for example (see above), and it seems to exercise a considerable charm over the minds of politicians, as well as pamphleteers. Greg Barker has recently published an article discussing employment opportunities in the environmental sector that not only carries the title 'Your Green Economy Needs You', but also remarks that 'Britain's green economy will need a massive injection of skills which could amount to a new "low-carbon army"',[4] a striking echo of the 'Carbon Army' which Simms discusses in both *A Green New Deal* and *The New Home Front*.[5]

As would be expected of a politician, this analogy is left vague in Mr Barker's article, but Simms goes beyond the rhetoric of previous nation-unifying conflicts to enumerate those points he finds attractive:

> In just 6 years from 1938 British homes cut their coal use by 11 million tonnes, a reduction of 25 per cent
>
> By April 1943, 31,000 tonnes of kitchen waste were being saved every week, enough to feed 210,000 pigs
>
> Food consumption fell 11 per cent by 1944 from before the war, but thanks to a scientifically planned national food policy, the population's health got better
>
> Scrap metal was saved at the rate of 110,000 tonnes per week
>
> Use of household electrical appliances dropped 82 per cent. A war on waste, new social norms and rationing helped general consumption fall 16 per cent (and more so at household level)
>
> Between 1938 and 1944 there was a 95 per cent drop in use of motor vehicles[6]

However, while such demand control was indeed tolerated during the war, the population expected this hardship to be of relatively short duration, and it may prove to be an uncertain guide to reactions to long-term regulation. As Simms notes in his earlier work, but omits to discuss in the more recent study:

Of course it has to be remembered that during World War II, restraining measures were accepted by the majority because there was a hope and expectation that this enforced frugality would end once the war was over. Fighting climate change, and coping with energy and food price rises and shortages, will be a battle with no imminent end in sight.[7]

In addition, Simms is well aware that the changes he proposes require a shift in lifestyle, and that there will be those that are relatively disadvantaged. So, in tandem with calling for the United Kingdom to be repowered with green energy (a process he quaintly refers to as a 'modernising' of its infrastructure), and, most importantly, to proceed with a 'rapid economic decarbonisation', he also insists on a move

> [...] to levels of economic equality comparable with that, say, of Denmark, [which] would create an economic safety net to buffer the process of change.[8]

This is partly a matter of the socialisation of an overall cost, but also rendering a reduction in living standards tolerable, because they are perceived as general and therefore equitable. Caroline Lucas, in her preface, notes

> [...] the importance of fairness in creating popular support for tough measures [...] rationing and conscription were introduced as much in response to popular pressure from below as it was to a desire for national controls from above.[9]

Simms expands this view, and finds further values in economic levelling:

> In more equal societies... reduced 'status anxiety' lowers the pressure for conspicuous consumption. As a result we may find ourselves both happier and less prone to consumerist behaviour.[10]

The Happiness Factor

Wartime policies, such as employment due to war production (*materiel* production at any cost, it should be noted), the employment of women and rationing, all 'significantly increased effective economic equality'.[11] Thus, although standards of living may fall in the green economy proposed by Simms and Lucas, this will be tolerable, in their view, because of the compensating comfort resulting from social engagement and a suspension of interpersonal competition:

More equal societies are less prone to 'keeping up with the Joneses', that negative cycle of conspicuous consumption linked to status competition, creating instead a positive cycle. Reduced environmental costs commensurate with lower consumption and lower social costs, aligned to greater income equality, then work to compensate for any loss of conventional GDP income arising from a drop in wasteful 'throw-away' over-consumption.[12]

In his earlier work Simms has referred to this as the 'new well-being', and criticised the 'conflation of a growing economy with rising well-being'.[13] Much of this line of thinking draws on the Happy Planet Index (HPI), also published by NEF, and in arguments now conveniently summarised in a book by Nic Marks, *The Happiness Manifesto*, which presents Costa Rica as a role model, since it has a high HPI index, but low per capita GDP and environmental impact.[14] In a parallel line of argument Simms, implicitly invoking the Easterlin Paradox, contends that the United Kingdom's 'sense of satisfaction with life has flat-lined' despite several decades of economic growth. This is interesting, but it is a contentious subject,[15] and even if we were to grant that the Easterlin Paradox has some substance, it would obviously be incorrect to reason that since increasing wealth is not perfectly correlated with increasing happiness, a decrease in wealth would therefore increase happiness either in general or in any particular country.

We can conclude that the argument Simms presents is self-consoling. Having admitted that the green economy will be poorer, he looks around for reasons to be unconcerned at the destruction of wealth implied by his policies, and produces the hope that poverty will or might have the 'unexpected, positive outcomes' he sees as arising from wartime austerity.[16] Certainly, there appears to be some evidence to suggest that average health improved during the war, but this is in comparison to the Depression years of the 1930s, and it is not clear that a reduction in living standards today would produce a similar improvement. After all, England is a densely populated society, which in 2009 had 398 people per square kilometre as against the global average of 48 (and under 100 people per square kilometre in Costa Rica).[17] It is a considerable stretch to believe that our perception of well-being would be enhanced by either increased exposure to pathogens, largely held at low levels in our food chain by intensive energy consumption in refrigeration and sterilisation, or reduced availability of contemporary medical treatments. Indeed Simms himself grants that 'one of the most fundamental questions for the transition to a low-carbon economy' is:

> …how to maintain the social contract – health and education services and security in retirement – when conventional growth becomes constrained.[18]

This is a striking admission, and in many respects preferable to the insincere or uninquisitive rhetoric of triple wins that we observed earlier. Simms is, it seems, well aware of the implications of his project, and concedes that a reduction in Gross Domestic Product may inhibit or even prevent the achievement of other social objectives:

> The challenge is multiple: to deliver a low-carbon, low material throughput economy; to increase resilience in the face of the potential for increasingly severe and often external shocks; to promote greater equality and social justice; to find a new, respectful environmental etiquette for our lives, and to maintain and enhance levels of well-being.[19]

Simms responds by looking to the wartime administration for inspiration, and what he observes is intense coercion:

> Change was not tentative and incremental, it was deliberately bold and visible. […] There was rationing, or the distribution of fair entitlements to available resources, and key goods. And there were taxes on luxury goods. Altogether this led to reductions in waste and domestic consumption. Crucially there was an active industrial policy and a major re-orientation of industrial priorities – it wasn't left to the whims of the market place or to 'nudges' from economic policy. Backing it all up was a major programme of War Savings in which people's money was invested in securing a better future for all.[20]

To his credit he appears to find this alarming, and when attempting to describe 'what a modern equivalent would look like' he is clearly drawn to a less authoritarian system, but honesty obliges him to admit that:

> Whilst the modern state is unlikely to be the sole architect, agent and judge of change, it would have to set the parameters for the delivery of key transition objectives, through a combination of local, community and private actors.

However, this is a transparent verbal evasion, a euphemism for microscopic central economic planning and a rigorous enforcement policy that bears down heavily on every level of society. The role of the state in Simms' projection would be pervasive and in all probability oppressive. It is worth reflecting on the degree to which such a role is also implied in the views of UNEP, the EU, or of OECD leaders such as Mr Cameron.

Concern as to the political structures likely to obtain if Simms' vision were to be realised are confirmed by the comfort he draws from the experience of Cuba, which he describes as the 'anti-model'. This is a country, Simms tells us, that as a result of economic embargoes has undergone many of the resource constrictions that threaten us, and is also regularly afflicted by extreme weather events, but has nevertheless not only survived, but actually prospered:

> By all accounts, Cuba should be a complete basket case – battered equally by the weather and its neighbouring superpower. In the face of all these challenges, why isn't Cuba on a par with some of the worst failed states in the world? Why does it not have shattered health and education systems? Why do its people not starve or suffer endemic malnutrition? The answer can be found in a rigid and centrally controlled economy, government planning, preparation and the fact that challenges were tackled courageously and imaginatively.[21]

Other perspectives on that country's history and current affairs are, of course, possible, but Simms appears unaware of them or unwilling to admit that they have any relevance, and that insouciance greatly undermines the credibility of his argument.

However, the relationship between these views and those found in the elliptical political utterances on the green economy, or the more elaborated but incomplete statements of UNEP, is important. Whereas Mr Cameron and the EU Commission explicitly promise continued economic prosperity as a result of their green policies, and UNEP does so in a more complicated and obscure fashion, NEF candidly admits that it cannot see such a pattern of sustainable development as being plausible, and that reductions in wealth are inevitable.

The gulf here is apparently large. It comes as a surprise, then, to find that the technical studies that underlie the policies of the EU, for example, are in fact closer to the Simms position than to the apparent optimism of Mr Verheugen, or the green hope of President Obama.

Chapter Three
Renewable Jobs in the EU

Introduction

The European Union's Renewable Energy Directive requires that 20 per cent of overall Final Energy Consumption (FEC) in the EU-27 should come from renewable sources by 2020, a target that covers transport, fuel and heating as well as electricity. This is an extraordinarily high level to achieve in a short period of time, and for some states their burden share implies high costs. Indeed, the United Kingdom, which has been allocated a target of 15 per cent of FEC by 2020, from the present level of just under two per cent, would, according to Department of Business Enterprise and Regulatory Reform (BERR) calculations leaked in 2008, shoulder around 40 per cent of the EU-wide costs,[1] a burden that is iniquitous and should be renegotiated.

However, belief in the potential for rapid and beneficial growth in the green sector is a consistent and often repeated element in official statements from the European Commission. For example, the recent communication to the European Parliament and the Council, 'Renewable Energy: Progressing towards the 2020 Target', issued on 31 January 2011, opens with the following sentence:

> Renewable energy is crucial to any move towards a low-carbon economy. It is also a key component of the EU energy strategy. The European industry leads global renewable energy technology development... employs 1.5 million people and by 2020 could employ a further 3 million.[2]

The prominence given to this confident statement is in itself highly significant, though its content is of course familiar from many other similar remarks. The foundations for this important sentence are provided in a footnote, which refers to a Commission study, *EmployRES,* and notes that these numbers are 'gross employment effects'. *EmployRES: The Impact of Renewable Energy Policy on Economic Growth and Employment in the European Union* (27 April 2009) is a study commissioned and funded by the European Commission's Directorate General of Energy and Transport (DG TREN), and was produced by six collaborating consultancies from Germany, The Netherlands, France, Austria, Switzerland and Lithuania.[3] The result is a substantial document (the summary alone is 27 pages long) and forms the major technical support for the EU's green economy agenda, and by implication for the hopes of individual member states.

However, its fundamental findings are by no means as reassuring as might be expected, certainly as might be expected from Commission citations, and *EmployRES* deserves to be much better known and more widely discussed, particularly in the UK, for which the study predicts a policy-induced brake on economic growth and net employment losses even in some of the most optimistic scenarios. Furthermore, the study finds that at best the EU's overall economic gains from the renewables policies are, to use *EmployRES's* own term, 'slight', and in any case almost entirely dependent on the EU maintaining a more than 50 per cent share of the global green technology market, and thus maintaining high levels of exports.

Not all these points are evident in the Summary, and it has to be noted that there is a tendency to omit detailed reporting of the more troubling findings in that overview document. Still more significantly, the Commission's use of the study in supporting claims such as that which opens the Communication quoted above is potentially deeply misleading. Bearing in mind the Commission's figure of 3 million jobs, and the footnote referring to 'gross' employment, we can turn to the first paragraph of the *EmployRES* summary:

> Improving current policies so that the target of 20 per cent RES in final energy consumption in 2020 can be achieved will provide a net effect of about 410,000 additional jobs and 0.24 per cent additional gross domestic product (GDP).[4]

The difference is immediately apparent. Whereas the Commission had reported a gross effect of three million additional jobs, *EmployRES* quite properly acknowledges that the net effect produces a much smaller number, since over two million jobs in other industries are destroyed as a result of the displacement and energy price increasing effects of the renewables policies. As we will see when reviewing the details of the study, even this modestly positive EU-wide effect is dependent on hopeful and arguably unrealistic assumptions, particularly with regard to exports, and glosses over local effects that are negative, in the UK for example. Before embarking on this summary fair warning should be given: even an abbreviated summary of *EmployRES*'s principal findings is necessarily intricate, and some readers may wish to move straight to this chapter's Conclusion, before examining the various charts displayed in the text.

EmployRES Methodology and Assumptions

The authors of *EmployRES* use an 'input-output' model (MULTIREG) to estimate the impacts of renewable energy sector development, itself predicted by another model (GREEN-X), on other economic sectors. These macro-economic impacts are examined via two independent models, NEMESIS and ASTRA, and the results compared. The authors describe this as 'the first study to assess the economic effects of supporting RES [renewable energy sources] in this detail, looking not only at jobs in the RES sector itself, but taking into account its impact on all sectors of the economy'.[5] In other words, the study attempts a rigorous investigation of both the gross and the net impacts of the policies. The authors write:

> Increased use of RES has various effects on the economy, some of which are positive in terms of employment and economic growth, while others are negative. This study presents both gross and net effects. Broadly speaking, gross effects include only the positive effects in RES and RES-related industries, while net effects are the sum of positive and negative effects. For the net effects, all relevant economic mechanisms are considered.

These mechanisms include:

> Increased investments, operation and maintenance costs and biomass fuel supply for RES
>
> Reduced investments, operation and maintenance costs in the conventional energy sector
>
> Fossil fuel imports and use avoided
>
> Increasing energy costs and their effects on the economy due to reduced competitiveness (industry) or reduced budgets for consumption (consumers and governments)
>
> Trade in RES technology and fuels among EU countries and with the rest of the world[6]

We should note in passing that assessing these mechanisms involves making numerous assumptions, for example with regard to current and future materials costs, fossil fuel costs, the performance of renewables, the integration costs for variable renewables, and many other factors, usually relying on industry sources. With regard to these matters we will, for the sake of the present argument, take *EmployRES* at face value, though it is as well to be aware that empirical experience has shown that industry estimates of such matters as system integration costs and wind power load factor onshore have proved to be unduly optimistic, with implications for the cost per kWh generated.[7] Such models, then, are inherently susceptible to error, and are therefore of indicative rather than precise value. This is a point to which we will return when evaluating the *EmployRES* findings and drawing conclusions from those results.

EmployRES also posits three policy scenarios:

1. No Policy for renewables support. In this scenario all current policies are abandoned.

2. Business as Usual (BAU). In this scenario the current (2009) renewables policies in the various EU states continue, but they are not augmented. This scenario is, the authors tell us, not adequate to meet the 2020 EU Renewables Directive, since it delivers 14 per cent of EU final energy consumption in 2020, and 17 per cent in 2030.

3. Accelerated Deployment Policies (ADP), these stronger support mechanisms delivering 20 per cent of EU FEC in 2020 and 30 per cent by 2030.[8]

These scenarios for renewables deployment are combined with three further scenarios describing the EU's share of the world market for renewable energy technologies:

1. Pessimistic Exports (PE). In this scenario the EU's market share falls from 69 per cent in 2009 to 31 per cent in 2030

2. Moderate Exports (ME). In this scenario the EU's share falls to 43 per cent.

3. Optimistic Exports (OE). In this scenario the EU's share falls only to 54 per cent.[9]

A great deal hinges on these market share scenarios, which are described in the following chart:

Figure 3:1
World market shares of the EU and the rest of the world (RoW)
in the global cost components of RES technologies
(weighted average of all technologies)

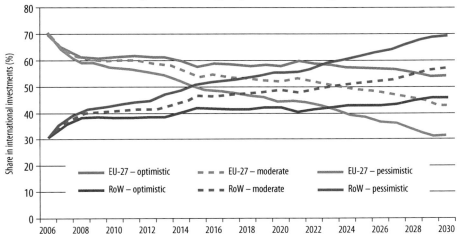

Source: *EmployRES*[10]

It is important to note that the empirical section of the chart shows that market share is declining, and all the predictive scenarios assume that this decline will continue, the variation between them being only in the rate of decline. This seems reasonable given the manifest comparative advantage enjoyed by China and India, amongst others, in certain areas of engineering, electronics, and manufacturing. The significance for net economic impact of a reduction in market share is twofold: firstly, if exports decline, then the EU loses the benefit of that income; secondly, the exports of the rest of the world will rise in part because of exports to the EU, therefore imposing an economic cost.

However, the matrix of possibilities considered only assesses the pessimistic export scenario in relation to the No Policy scenario. In other words, the situations in which the EU maintains current policies (BAU), or adopts augmented policies (ADP) are only assessed in relation to moderate and optimistic export scenarios (ME and OE).

That is to say *EmployRES* does not consider the scenario combination in which the EU has aggressive renewables support mechanisms, at high cost to the consumer, but only a very low share of the international export market. The authors of *EmployRES,* with some justification, consider this nightmare scenario to be unlikely; nevertheless, it would have been interesting to see the results for the pessimistic combinations.[11] For the sake of clarity we can represent this incomplete engagement in a table:

Table 3:1

Matrix of policy scenarios, export assumptions and macro-economic models. 'Yes' indicates that EmployRES publishes findings relating to the relevant combination; 'No' indicates that the combination is not considered in the study

	Macro-economic Model: ASTRA			Macro-economic Model: NEMESIS		
	No Policy	Business as Usual	Accelerated Deployment Policies	No Policy	Business as Usual	Accelerated Deployment Policies
Pessimistic Exports	Yes	No	No	Yes	No	No
Moderate Exports	Yes	Yes	Yes	Yes	Yes	Yes
Optimistic Exports	Yes	Yes	Yes	Yes	Yes	Yes

EmployRES Findings: Employment Effects of Renewable Energy Policies

The results generated by *EmployRES's* method are displayed in over 60 charts, several of which appear only in the Summary and many of which appear only in the main text. A complete picture of the study's findings requires reference to both documents. We will begin with a consideration of the study's findings in relation to the EU considered as a whole, and then move to the discussion of the effect on individual member states.

(i) Employment Effects on the EU as a Whole

The following pair of charts describes the gross and net employment effects of RES policies in the overall EU economy, the gross effects being calculated from NEMESIS, and the net effects from both NEMESIS and ASTRA:

Figure 3:2
Employment effects by 2020 in the EU-27, showing the gross increase in jobs (1,000s) in the Renewable Energy Sources (RES) sector (left) and the net increase in jobs in the whole economy as a result of RES policies (right)

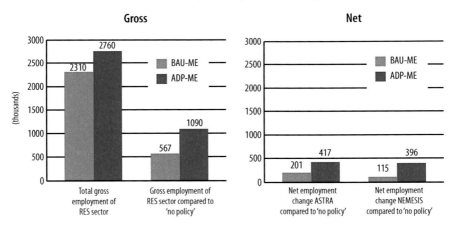

Source: *EmployRES*[12]

Of the left-hand chart the accompanying text in the Summary remarks:

> Total gross employment in the RES sector in the EU-27 in 2020 will amount to 2.3 million people under the BAU-ME scenario and 2.8 million under the ADP-ME scenario. Compared to the hypothetical scenario in which all RES support policies are abandoned, the additional gross employment due to RES policies amounts to 0.6 million people for the BAU-ME scenario and 1.1 million people for the ADP-ME scenario. Total gross employment in the RES sector may increase by up to 3.4 million people by 2030 if there is an accelerated deployment policy combined with optimistic export expectations (ADP-OE).[13]

These are large numbers, and in the main text the authors observe that on this view the renewable energy industry would 'become one of the very important sectors in terms of employment in Europe'.[14] While such gross figures are of limited value in many respects, they do shed light on the rebalancing of the EU economies that is implicit in a subsidised and target-driven transition to renewables. A government-mandated employee base on this scale has significant implications for energy prices, and thus for net economic effects in the longer term if these jobs are to be maintained permanently at non-market wages. Indeed, the marginal net employment effects reported by *EmployRES* in the right-hand chart for both the ASTRA and the NEMESIS macroeconomic models above confirm the view that the cost of supporting renewables causes significant contraction in other parts of the economy due to, in the words of the study itself, 'increasing energy costs and their effects on the economy due to reduced competitiveness (industry) or reduced budgets for consumption (consumers and governments)'.[15] Commenting on this suppressive effect, the authors write:

> Sectors losing employment would suffer from the higher energy expenditures of households, the higher sectoral elasticities in response to higher goods prices driven by energy cost increases and the prevailing budget constraint of households. Examples would be the trade and retail sector as well as the hotels and restaurant sector.[16]

The effect on energy-intensive users, the steel and chemicals industries for example, should have been mentioned here, but it is useful to be reminded that higher energy prices have an important indirect impact on service industries.

Furthermore, as noted, *EmployRES* only reports findings relating to Moderate Export (ME) and Optimistic Export (OE) scenarios. The net employment effect in ADP-ME, where the EU holds over 40 per cent of the global export market in renewable energy technologies, is only weakly positive, and the study itself refers to the effect as 'slight'.[17] We are left to guess what sort of effect would obtain in the pessimistic scenarios, where the EU holds only just over 30 per cent of the global export market in renewable energy technologies. These findings are sobering. Even on the assumption that the EU retains a substantial share of the global export market, the net employment effects of the renewable energy policies are revealed by *EmployRES* to be weak, and we cannot avoid the inference that they are, in all probability, fragile.

(ii) Employment Effects on Individual Member States

As might be expected, these effects are not evenly distributed across the EU-27, and the study helpfully provides gross and net employment effects analysed by member state. The following chart, which appears in both the summary and main text, represents the gross employment impacts of the Additional Deployment scenario in conjunction with the Moderate Export scenario, employing the NEMESIS model.

Figure 3:3
Relative and absolute differences in employment between Accelerated Deployment Policies and Moderate Exports (ADP-ME) scenario and the No Policy scenario for the 2020, by countries and in relation to total employment in 2007

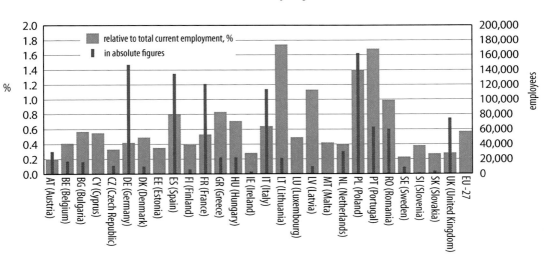

Source: *EmployRES*[18]

Percentage change is indicated by the grey bars and the left-hand axis, and absolute numbers by the black bars and the right-hand axis. It is important to note that even the gross effects in comparison to No Policy are truly marginal for many member states. Gross employment figures analysed by member state are provided for no other scenario than ADP-ME, which hampers consideration of the net effects to be discussed later, particularly in relation to the United Kingdom.

However, this chart can be taken as indicating the approximate scales to be considered.

The net employment impacts on the EU-27 members are described in seven charts, three relating to the NEMESIS model, and four to ASTRA. The policy and export scenarios considered are Business as Usual – Moderate Exports (BAU-ME); Business as Usual – Optimistic Exports (BAU-OE); Accelerated Deployment Policies – Moderate Exports (ADP-ME); and Accelerate Deployment Policies – Optimistic Exports (ADP-OE). One scenario is omitted, NEMESIS ADP-OE, perhaps in error.[19] As noted before, no pessimistic scenarios are considered, on the grounds that they are of lower probability. For clarity we can represent these combinations in a matrix.

Table 3:2

Matrix of policy scenarios, export assumptions and macro-economic models. 'Yes' indicates that EmployRES publishes findings relating to the relevant combination; 'No' indicates that the combination is not considered in the study

	Macro-economic Model: ASTRA		Macro-economic Model: NEMESIS	
	Business as Usual v. No Policy	Accelerated Deployment Policies v. No Policies	Business as Usual v. No Policies	Accelerated Deployment Policies v. No Policies
Pessimistic Exports	No	No	No	No
Moderate Exports	Yes	Yes	Yes	Yes
Optimistic Exports	Yes	Yes	Yes	No

Though the scenario coverage is incomplete, the *EmployRES* analysis of employment effects delivers important results, and the failure of the Commission to publicise them in order to facilitate balanced discussion of the wisdom of the proposed policies is regrettable.

In only one of the seven scenario combinations considered is there a net positive employment gain for the United Kingdom, namely NEMESIS, ADP-ME (the gross scenario for which is described above), which shows a gain of approximately 2,500 jobs. In all other scenarios charted, the UK, alone of the EU-27, records net negative employment, ranging from a net loss of over 10,000 jobs to a net loss of over 30,000 jobs. Bearing in mind the scale of the gross job creation shown for the ADP-ME scenario (approximately 70,000 jobs), it is clear that the economic impact of the EU renewables policies on the United Kingdom is significantly negative.

In the interests of concision, we will reproduce only four of the scenarios considered for both the NEMESIS and ASTRA macroeconomic models.

Figure 3:4
NEMESIS: Change in employment: Business as Usual and Optimistic Exports (BAU-OE) compared to No Policy

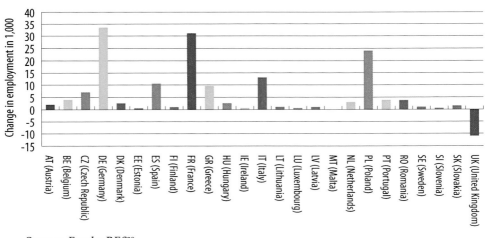

Source: *EmployRES*[20]

Few EU states exhibit even modest net gains, most being marginal. The UK suffers a net loss of over 10,000 jobs.

Figure 3:5
**NEMESIS: Changes in employment: Accelerated Deployment
Policies and Moderate Exports (ADP-ME) compared to No Policy**

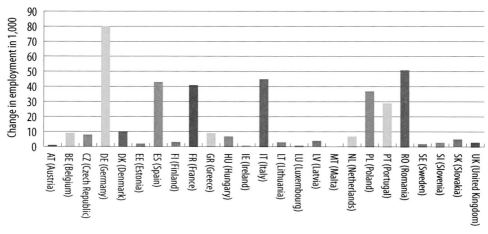

Source: *EmployRES*[21]

This is the sole chart displayed in which the UK has as net positive
employment gain, which can be estimated at approximately 2,500
jobs. The gains in competitor states such as Germany, France and
Spain are significantly higher.

The findings in the ASTRA model are still less encouraging. Again
we will consider only the optimistic export scenarios:

Figure 3:6
ASTRA: Change in employment: Business as Usual and Optimistic Exports (BAU-OE) compared to No Policy, 2020

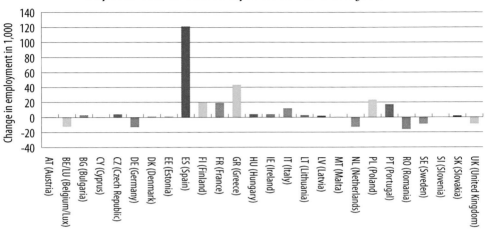

Source: *EmployRES*[22]

In this scenario many EU states suffer net negative employment effects, and the gulf between the winners and losers appears to be greater than before.

The United Kingdom experiences net negative job effects of around 30,000 jobs, with the Netherlands also seriously affected. Other countries such as Spain fare better.

EmployRES Findings: GDP Effects of Renewable Energy Policies

(iii) GDP Effects on the EU as a Whole

We have already observed the headline figure for the EU offered in the *EmployRES* summary of 0.24 per cent of net additional GDP as compared to the no-policy scenario in which current renewable energy policies are abandoned.[23] It is interesting to further note that the Summary also reports that, under the NEMESIS model: 'Assuming an accelerated deployment policy combined with optimistic export expectations (ADP-OE) *net additional* GDP compared to the no-policy scenario would amount to 0.44 per cent of GDP in 2030'.[24] Reference to the useful chart comparing Gross Value Added and net GDP changes confirms this point:

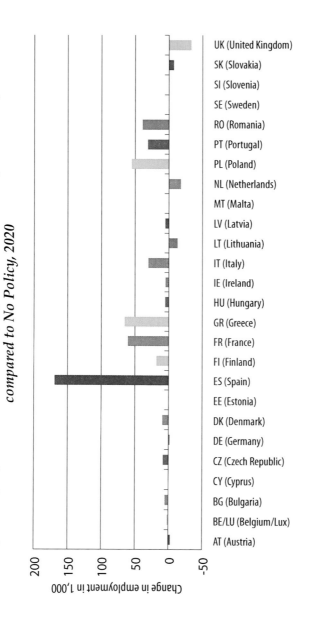

Figure 3:7

ASTRA: Change in Employment: Accelerated Deployment Policies and Optimistic Exports (ADP-OE) compared to No Policy, 2020

Source: *EmployRES*[25]

Figure 3:8
Economic growth effects by 2020 in the EU-27 showing the gross value added of the Renewable Energy Sources (RES) sector in the NEMESIS model (left) and the net GDP impact of RES policies (right) in both NEMESIS and ASTRA, both as a ratio of GDP. Both Business as Usual and Accelerated Deployment Policies are considered in relation to the Moderate Export scenario.

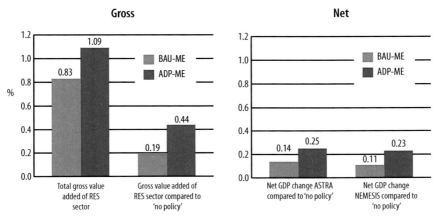

Source: *EmployRES*, p.6

This would appear to be a small gain, well within the measuring error, and hardly proportionate to the economic and technological risks involved. It is troubling that Pessimistic Export scenarios are not assessed.

(iv) GDP Effects on Individual Member States

Viewed from the perspective of individual member states, the effects on GDP are also discouraging. The scenario and assumption combinations considered are the same as for net employment, though on this occasion the NEMESIS ADP-OE combination is present. All the results for the NEMESIS set show that the UK sees relative economic contraction, while the results for the ASTRA model show slight growth.

We can illustrate this with two charts for the most optimistic export scenarios. NEMESIS (Figure 3:9) indicates that while most EU states experience slight GDP growth as a result of the renewables policies, the UK experiences relative contraction, and ASTRA (Figure 3:10)

35

shows that GDP growth is modest even under an optimistic export scenario.

Figure 3:9
NEMESIS. Change in GDP: Accelerated Deployment Policies and Optimistic Exports (ADP-OE) compared to No Policy, 2020

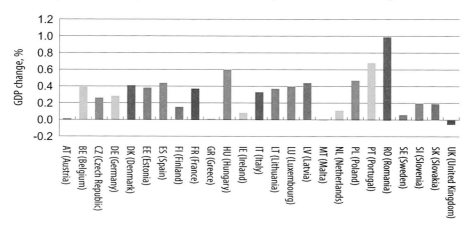

Source: *EmployRES.*[26]

Figure 3:10
ASTRA: Change in GDP: Accelerated Deployment Policies and Optimistic Exports (ADP-OE) compared to No Policy, 2020

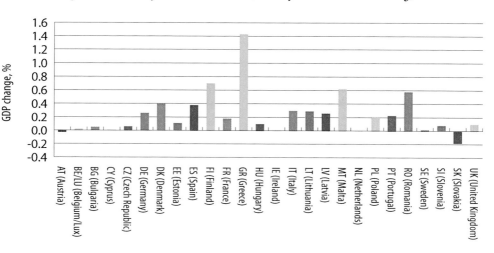

Source: *EmployRES.*[27]

Conclusion

In summary, the *EmployRES* modelling exercise shows that even in scenarios assuming optimistic European dominance of the world market for renewable energy technology, the net employment effect for the United Kingdom will be negative, as rising energy prices cause economic contraction in other parts of the economy. Net employment effects in other EU states are varied, and clearly heavily dependent on optimistic assumptions with regard to exports to other EU states and the rest of the world.[28]

Effects on GDP are more mixed, with the NEMESIS and ASTRA models at slight variance. The authors themselves note that the results show only 'slight' overall growth even in scenarios making optimistic assumptions with regard to renewable energy equipment exports (0.25–0.26 per cent in 2020).[29] For some states, notably the UK, the prospect is for either relative economic contraction or only modest growth. The unconsidered pessimistic export scenarios would presumably produce results that were still less encouraging.

We can summarise these findings in the following table, where the numerical values reported are measured optically from the charts, and are therefore approximate:

Table 3:3

Summary of results for EmployRES's modelling of net employment and GDP effects of renewable energy policies in the United Kingdom to 2020, compared to No Policies

	Macro-economic Model: ASTRA		Macro-economic Model: NEMESIS	
	Business as Usual v. No Policy	Accelerated Deployment Policies v. No Policies	Business as Usual v. No Policies	Accelerated Deployment Policies v. No Policies
Pessimistic Exports	Not considered	Not considered	Not considered	Not considered
Moderate Exports	Jobs: –10,000 GDP: +0.07%	Jobs: –31,000 GDP: +0.01%	Jobs: –11,000 GDP: –0.01%	Jobs: +2,500 GDP: –0.03%
Optimistic Exports	Jobs: –10,000 GDP: +0.07%	Jobs: –31,000 GDP: +0.1%	Jobs: –11,000 GDP: –0.01%	Jobs: Not considered GDP: –0.03%

There is little difference between the various outcomes, and even such rewards as obtained under the optimistic scenarios do not seem commensurate to either the scale of the endeavour or the implied risks. Put another way, if economic reform on the scale proposed is to be undertaken, with all its attendant technological, social and financial dangers, there should be a potential for significant gains to justify the adventure. However, the *EmployRES* study suggests that though the hazard is large, the prizes are minor at best.

Concerns of this type are compounded by the likelihood of opportunity cost resulting from the macroeconomic inflexibility entailed by commitment to the renewables policies: in other words, the technological inventions and innovations necessarily forgone because of the mandated pursuit of currently available renewable energy sources.

Finally, as noted earlier, complex models of this kind must embed many assumptions about technology cost and performance, and there is every reason to have concerns about these matters. Empirical data in the UK, for example, shows that over the last decade the wind industry overestimated its likely onshore performance, with load factor since 2003 ranging between 23 and 28 per cent, rather than the 30 per cent anticipated. Indeed, year-on-year variation in load factor is now known to be highly significant, with the difference between that achieved onshore in 2009 and that of 2010 being as great as five percentage points. Similar results have been observed in Ireland, where Eirgrid reports that onshore load factor[30] in 2010 was approximately 23.5 per cent, as compared to 31 per cent in the previous year, and an average figure of 32.3 per cent for the years 2002 to 2009.[31] Empirical experience has also revealed strongly correlated low wind power output over large geographical areas at times of low and very low ambient temperature, confirming theoretical calculations that, regardless of the size of the wind fleet, the conventional generation sector can never be smaller than peak load plus a margin of ten percent.[32] Such findings have resulted in renewed concern over the system cost of integrating large quantities of uncontrollable renewables while simultaneously maintaining a robustly reliable electricity system. The scale of the benefits observed in *EmployRES* is not sufficient to give comfort in the presence of such technological uncertainties.

Chapter Four

The Low-carbon Economy: A Photofit

Looking back over the European Union's *EmployRES* study, the visions of NEF and UNEP, and the elliptical but deeply significant political remarks with which this section started, we are now in a position to produce a composite image of the low-carbon economy as it is envisaged in a wide variety of sources.

At the centre lies the concept of state control. No attempt to realise any of the various visions under consideration can be made without coercion of energy producers and consumers, and in order to make progress the state must purchase the compliance of producers by disarming consumers. The employment gains projected by proponents of the low-carbon economy entail an enlargement of the energy sector, and, if wages are to remain high, a significant increase in price to consumers, whose purchasing power will decline. This will entail a contraction of the residual economy, even if the total size of the economy remains the same or grows. In effect, the low-carbon economy is dependent on a state-mandated transfer of wealth from consumers and non-energy producers to the energy sector, which together with its governmental sponsor will come to dominate the societies that it supplies.

Political leaders, and politicised bodies such as UNEP, prefer to suggest that this is compatible with conventional definitions of high standards of living for the general population, but more candid thinkers such as Simms envisage poorer (though happier) societies resulting from this transition. Paradoxically, technical analysis produced for governmental bodies tends to suggest that this latter vision is more probable. However, Simms and UNEP are probably mistaken in thinking that more equal societies will result; gradients of power will exist between favoured industries, the low-carbon sector, and all others, and it would be surprising if these did not translate into differentials of well-being.

States that succeed in dominating the export markets for renewables will be the net beneficiaries of wealth transfers from states that mandate the adoption of renewables. The EU currently accounts for the vast majority of the global market, but anticipates that it will lose market share, making overall benefits to the EU marginal.

Whether or not we find this photofit image attractive, we are now in a position to ask some of the questions sketched in the introduction. What will it be like to try realise the low-carbon economy? What are the risks of the attempt? How long will it take? Assuming that it is realisable, at least substantially, will it be stable? Answers to some of these questions can be found by reflecting on three areas of practical experience. Firstly, the Rooseveltian New Deal, which is the inspirational root of much green economic thinking. Secondly, the United Kingdom's Groundnut Scheme, a key example of a major Western economy attempting to run a global-scale business enterprise. And thirdly, the evidence that can be derived from state mandated renewables growth in the two leading examples of that endeavour, Germany and Spain.

PART TWO

From the Archives

Chapter Five

The Green Deal and the New Deal[1]

The Green Deal

On 21 September 2010 the Department of Energy and Climate Change announced its Green Deal in a statement headed with the promise of 'Up to a quarter of a million jobs by 2030', and concluded with a statement by the Secretary of State, Chris Huhne MP:

> The Green Deal is a massive new business opportunity which has the potential to support up to a quarter of a million jobs as part of our third industrial revolution.[2]

Such claims would be a heavy burden for any policy programme to sustain, even conceptually, and when the offering is as far from rocket science as insulation and energy saving there is a risk of inadvertent comedy. Nevertheless, the goals are worthy, though the details of the scheme may give cause for concern:

> The Green Deal will be a new and radical way of making energy efficiency affordable for all, whether people own or rent their property. The upfront finance will be attached to the building's energy meter. People can pay back over time with the repayments less than the savings on bills, meaning many benefits from day one. It will help save carbon, energy and money off fuel bills.[3]

In public speeches, Minister of State for Climate Change Greg Barker has referred to the Green Deal as a 'flagship policy' for the Coalition Government,[4] and it would certainly appear to be high on the list of priorities, forming part of the December 2010 Energy Bill.[5] However, it seems to be a clear case where potential problems have been brushed aside by a department that seems more attached to the gestural value of the announcement than to the content. For example, there are major concerns about the difficulty of leaving properties with a charge that may impede mortgage lending. Furthermore, there is the clear risk that the devices may not deliver the savings expected, thus imposing the cost of unsaved fuel on the homeowner in addition to the cost of the failed equipment. When pressed on this latter point, Mr Barker responded that good certification was critical to the plan.[6] While doubtless desirable, such an oversight programme must judge installations on industry best practice, which is itself prone to error.

Empirical evidence shows that this is no idle point. The classic instance of this in relation to domestic heating equipment is the Japanese 'Solar Tragedy', where government-driven, rapid growth in the solar thermal heating sector resulted in the installation of sub-optimal technology, consequent consumer disenchantment, and a collapse in the annually installed capacity of that technology (nearly 2.75 million square metres in 1980, but only 0.25 million square metres today). The Japanese solar thermal market has yet to recover, in spite of a return to higher oil prices, as can be seen in the following chart:

Figure 5:1
The annually installed capacity (m²) of solar thermal technology in Japan, charted against oil price

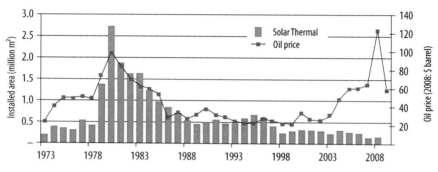

Source: Redrawn from 'Japan: New Policies to Spark Growth?'[7]

Arguably, Japan currently has a weaker solar thermal industry than it would have had without any state pressure.

The New Deal

Enthusiasm for such programmes as the Green Deal persists in spite of concerns that are not only theoretically well-formed but also supported by solid empirical foundations, and it does so for because the theatrical value of government intervention is as important as fundamental viability. The gestural combinations are numerous and potent, ranging from obvious correlations with deep-green labour union-oriented positions such as the Campaign against Climate Change to the echoes of President Obama's green stimulus package, with its emphasis on 'weatherization'. Most obviously of all, there is the New Deal itself. Indeed, there is every reason for thinking that Roosevelt's three-term presidency and its handling of the Depression has a direct and formative influence on a broad swath of current politics, an influence that is as much emotional as it is technical. It is interesting to reflect that many of Mr Cameron's generation in the United Kingdom were taught this period of American history at O-Level, as I was, and many others continue to absorb the implicit views and romantic aura of the period through set texts such as Steinbeck's *Of Mice and Men* and *The Grapes of Wrath*. Indeed, there can be few economic phenomena with so deep and engaging a gallery of representations in high and popular culture, ranging from the austere and wrenching documentary text and photography of James Agee and Walker Evans' *Let Us Now Praise Famous Men* to the heart-warming cinema of *It's a Wonderful Life*.

The relevance of this background to the low-carbon project goes well beyond a generalised precedent, though this is crucially important, since it is reinforced by the fact that the Tennessee Valley Authority, as every schoolchild was taught, built dams and engaged in a scheme of rural electrification that lifted the affected areas out of ages that were quite literally dark. Subsequently, those pupils have learned that this energy was renewable, a backwards reflection that can only combine fruitfully with the positive reputation of Roosevelt and his legislative programme. The trope of the Green Deal is all the more powerful since with hindsight the New Deal seems green before its time.

This is hardly surprising, and the impact of the New Deal on political discourse in the United Kingdom has been powerful from the outset, with Lloyd George attempting to launch a 'New Deal for Britain' in January 1935, a venture the failure of which had more to do with the wizard's declining magic than any intrinsic lack of interest. The impact of Roosevelt's New Deal on British politics in the 1930s was subtle rather than dramatic, and it rendered the public still more receptive to the suggestions that Beveridge would later make. Indeed, as one later historian has remarked, the Beveridge report did not change minds but confirmed existing opinions,[8] and while the sources of this consensus were numerous, Roosevelt's confident example was prominent amongst them. Even today, the New Deal continues to have a role of this kind in British politics, where it has been so fully digested and absorbed into the fabric of thinking about state intervention in social and economic affairs that even brief references can be used in the hope of invoking spontaneous assent to any related proposition. Examples are not hard to find, but a prominent illustration can be found in Gordon Brown's attempt to reform the labour market in the United Kingdom, introduced in January 1998, and explicitly and repetitively badged as a 'New Deal'.[9] Similarly, Mr Brown's recession-buffering policies were announced in January 2009 and presented to the press 'as a modern reworking of Roosevelt's New Deal',[10] the expectation clearly being that such an association is a charm to ward off criticism. To a large degree this confidence is justified: the wisdom and effectiveness of the New Deal is simply accepted, and it is therefore an immensely useful anchoring point for any novel proposition.

The Green New Deal

Co-options can be more or less thoroughly worked through, with political references tending to be glancing and casual, though drawing strength from roots in analyses that are much more deeply engaged. The Coalition's 'Green Deal' may be a suasive echo of the standard type, but it has a close relationship, perhaps even a direct one,[11] with a document published in 2008 by the New Economics Foundation on behalf of the Green New Deal Group, an organisation that relies on a detailed comparison between our present circumstances and the Great Depression.

Figure 5:2
Logos of the New Deal's Civilian Conservation Corp, 1933–42; and the Green New Deal Group, 2008

Since we are told that 'the views and recommendations of the report are those of the group writing in their individual capacities', there is some point in remarking that the Green New Deal Group comprises: Larry Elliott, Economics Editor of the *Guardian*; Colin Hines, Co-Director of Finance for the Future, and former head of Greenpeace International's Economics Unit; Tony Juniper, former Director of Friends of the Earth; Jeremy Leggett, founder and Chairman of the energy company Solar Century and the charity SolarAid; Caroline Lucas, Green Party (then MEP now MP); Richard Murphy, Co-Director of Finance for the Future and Director, Tax Research LLP; Ann Pettifor, former head of the Jubilee 2000 debt relief campaign, Campaign Director of Operation Noah; Charles Secrett, Adviser on Sustainable Development, and former Director of Friends of the Earth; and Andrew Simms, Policy Director, NEF (the New Economics Foundation). This is an interesting, even an important, group of individuals, several of them being both prominent and influential in the development of the environmental movement in the UK. It seems safe to take their views as being representative of a broad band of committed activist opinion, and of the occasional views of the general public. The group's overall standpoint is neatly summarised in the opening statement of the text:

> The global economy is facing a 'triple crunch'. It is a combination of a credit-fuelled financial crisis, accelerating climate change and soaring energy prices underpinned by an encroaching peak in oil production. These three overlapping events threaten to develop into a perfect storm, the like of which has not been seen since the Great Depression. To help prevent this from happening we are proposing a Green New Deal.[12]

These three concerns—economics, climate change, and resource erosion—are arranged around a central view that binds them together and suggests a network of integral, reciprocally reinforcing relations between them:

> The triple crunch of financial meltdown, climate change and 'peak oil' has its origins firmly rooted in the current model of globalisation. Financial deregulation has facilitated the creation of almost limitless credit. With this credit boom have come irresponsible and often fraudulent patterns of lending, creating inflated bubbles in assets such as property, and powering environmentally unsustainable consumption.[13]

In short, this is an anti-globalist position. While welding such varied and possibly disjunct views together in a diagnosis is simple, to do so in a policy recommendation requires further justification, and this is where Roosevelt's policies show their worth:

> Drawing our inspiration from Franklin D. Roosevelt's courageous programme launched in the wake of the Great Crash of 1929, we believe that a positive course of action can pull the world back from economic and environmental meltdown. The Green New Deal that we are proposing consists of two main strands. First, it outlines a structural transformation of the regulation of national and international financial systems, and major changes to taxation systems. And, second, it calls for a sustained programme to invest in and deploy energy conservation and renewable energies, coupled with effective demand management.[14]

In such an argument the New Deal is not only an example of successful, integrated economic planning, but comes with pre-established good public standing and western democratic credentials, a powerful and, in many ways, persuasive combination.

Under this governing statement the authors arrange their sub-policies, each one connected to the central principles and to each other. There is to be a crash programme costing £50 billion a year to make 'every building a power station'. While it is not clear whether those employed in this activity will be engaged by the private or public sector, the fact that the study refers to this workforce as a 'carbon army' suggests that it will be under state control to a substantial degree. The example of Germany is given as a country currently benefiting from the 'boom in "green collar" jobs' now employing 250,000 people.[15] Funding for this ambitious programme will, at least initially, come from carbon taxes, which will also provide social policies to protect the poor from rising costs. Further government coercion in the financial sector will develop a 'wide-ranging package of other financial innovations and incentives' to marshal the funds that must be spent. This is justified on the grounds that:

> The science and technology needed to power an energy-and-transport revolution are already in place. But at present the funds to propel the latest advances into full-scale development are not.[16]

In other words, rather than attribute slow adoption of renewables to fundamental immaturity or weakness, the authors assume that it is sheer market perversity or blindness that holds them back, an odd assumption given that elsewhere the investment markets are supposed to be concerned only with profit. Indeed, a large part of the study's recommendations are concerned with financial institutions, and include the provision of low interest loans, the forced demerger of banking and finance groups along the lines of Roosevelt's Glass-Steagall legislation,[17] the strict regulation of derivative instruments, and a clamp-down on tax havens. The populist tone here is strong, with the authors promising that their measures will return finance to its 'role as servant, not master, of the global economy' through the 'restoration of policy autonomy to democratic governments' and the 'reintroduction of capital controls'.[18]

Was Roosevelt right?

It is interesting to imagine this outline in the absence of any reference to the New Deal. Without Roosevelt's prestige, and the reassurance of the fireside chat, these proposals would be harder to distinguish from the strident and extremist schemes contained in the pamphlets of the Campaign against Climate Change. But most important of all, the New Deal is a model that pre-empts accusations of impracticality. FDR, as everyone thinks they know, was successful.

In fact, there is a substantial and growing body of empirical analytic economic history that suggests that we may have misunderstood both the causes of the Great Depression and the efficacy of the New Deal as a remedy. The received wisdom, on which much of what we have described above is based, starts in the 1930s itself, finds full form in classic histories such as William Leuchtenberg's *Franklin D. Roosevelt and the New Deal, 1932–1940*,[19] and Friedman and Schwartz's *Monetary History of the United States*.[20] It remains vigorous and current in standard popular sources such as Rauchway's recent and entertaining *The Great Depression and the New Deal*.[21]

On this view the Great Depression started in 1929 as a 'common or garden' recession that was exacerbated by President Hoover's determination to retain balanced budgets and refrain from intervention. The rapidly worsening situation was compounded by undersupply of money from the Federal Reserve – this is Friedman and Schwartz's point – that ignited a full scale deflationary economic crisis resulting in bank collapses and widespread unemployment. On this view, Hoover's continued refusal to intervene resulted in the onset of the Great Depression. As one contemporary claimed, Hoover 'clung to the time-worn Republican policy: to do nothing, and when the pressure becomes irresistible, to do as little as possible'.[22] This situation was only rescued by the election of Roosevelt, whose landslide victory was won, as Rauchway puts it, 'on the hope that he, as Hoover on principle would not, might bring relief to ordinary Americans'. Roosevelt rapidly stabilised the banks and embarked on the New Deal, a programme of unprecedented government spending that created employment and steadily restored the economy to growth. Even with these measures, the severity of the Depression meant that full recovery did not occur until the American economy was galvanised by the need to manufacture war *materiel*.[23]

This narrative is a vindication of interventionist economics, and for some, such as Paul Krugman, the events of the late 1920s and early 1930s are an ethical watershed, and a touchstone for understanding the fundamental division of opinion animating subsequent political disagreement:

> One side of American politics considers the modern welfare state – a private-enterprise economy, but one in which society's winners are taxed to pay for a social safety net – morally superior to the capitalism red in tooth and claw we had before the New Deal. It's only right, this side believes, for the affluent to help the less fortunate.
>
> The other side believes that people have a right to keep what they earn, and that taxing them to support others, no matter how needy, amounts to theft... There's no middle ground between these views.[24]

With so much depending on the prevailing assessment of the New Deal, any proposed revisions are bound to be controversial, though correspondingly important and, due to emotional inertia, slow to move into the mainstream consciousness. In fact, while some of this new thinking has become topical in the light of President Obama's stimulus policies, the research on the fundamental economic data relevant to reassessing the causes of the Depression and Roosevelt's policies has been in process for some considerable time. Cole and Ohanian's seminal papers, 'The Great Depression in the United States from a Neoclassical Perspective', and 'New Deal Policies and the Persistence of the Great Depression: A General Equilibrium Analysis', were published in 1999 and 2004 respectively.[25] Taken together with Ohanian's more recent paper, 'What – or Who – Started the Great Depression?',[26] the data and analysis presented in this view suggests that the Depression had different causes than previously thought, that it lasted much longer, and that, in spite of local alleviations, Roosevelt's efforts were actually counterproductive.[27]

Firstly, Ohanian notes that the recession under Hoover was no minor matter, and that far from being a common or garden recession, an industrial depression began abruptly and severely in late 1929, before the banking crisis of early 1930. Output and hours worked fell by 20 per cent by June 1930, and were down 40 per cent by the autumn of 1931.

Figure 5:3
United States manufacturing hours and output. September 1929 = 100.

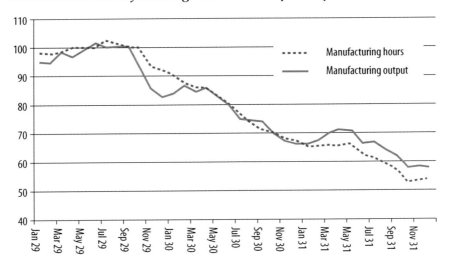

Source: 'What – or Who – Started the Great Depression?'[28]

Ohanian observes that the situation in the agricultural sector, which had about the same share of employment as industry in 1929, was different, with hours worked and output both rising slightly between 1929 and 1931. The explanation of this remarkable difference,

Ohanian suggests, is that President Hoover intervened in the industrial sector to discourage lay-offs (he encouraged work sharing, which reduced productivity) and protect wages, thinking that higher wages would stimulate the economy through spending. As a consequence, real wages rose in industry during the onset of the depression, while in the agricultural sector, by contrast, they fell. Furthermore, Hoover fostered Trade Associations in order to reduce competition. In other words, while Hoover was in many other respects a *laissez-faire* president, during the Great Depression he was not so. On the basis of economic modelling, Ohanian suggests that Hoover's interventions were responsible for roughly two-thirds of the depth of what was in fact a deep, immediate and sectorally asymmetric, depression.

In dealing with the banking crisis, Roosevelt was successful in restoring confidence, and deflationary problems were rapidly corrected, but in other respects he continued with policies that, like Hoover's, distorted labour markets. Indeed, prominent New Deal proponents, such as General Johnson, reasoned from their own experience as economic planners in the First World War, which had seen a period of economic expansion, that a suspension of competition would cause wages and output to rise. However, just as Hoover's policies had caused the Great Depression, Roosevelt's manipulation of wages and employment levels caused it to persist, with the result that the American economy, which was fundamentally strong and should have returned rapidly to growth, reaching trend by 1936, did not recover until the Second World War. In fact, employment and output remained well below their 1929 levels in 1939, and Ohanian and Cole remark that the empirical data suggests that there was almost no recovery during the New Deal period, with real output being 25 to 30 per cent below trend in the late 1930s.

Ohanian singles out for particular criticism the National Industrial Recovery Act (NIRA) of 1933 (declared unconstitutional in 1935) and its descendent legislation, noting that it encouraged explicit collusion, minimum prices, production quotas and collective bargaining, with the consequence that wages rose as the monopoly rents were shared with labour. Cole and Ohanian conclude that the NIRA's effects depressed employment, output and investment in the sectors covered by the Act.

Figure 5:4
***Predicted and actual output in 1929–1939. Detrended levels,
with initial capital stock in the model equal to actual
capital stock in 1929.***

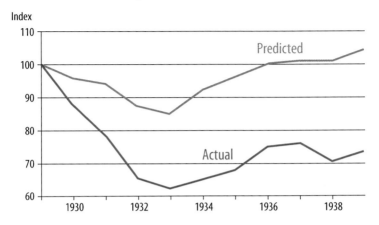

Source: 'The Great Depression in the United States from a Neoclassical Perspective'.[29]

There is no consensus yet on these views, but Cole and Ohanian's work is empirical, rigorous and, in many respects, persuasive. At the least it shows that an uncritical attitude towards the New Deal is untenable; there are now reasonable doubts regarding the traditional account of Roosevelt's presidency. Current policy systems grounded in the view that the New Deal was a successful interventionist economic plan should, consequently, be treated with considerable caution.

Chapter Six

'Operation Groundnuts'

Introduction

While new controversies emerge about the achievements of the New Deal, there is no such disagreement about a large-scale initiative brought forward by the post-war British Government some dozen years afterwards, and certainly no likelihood of anyone in today's British Government making a New Deal-style allusion to it. The East African Groundnut Scheme, to give it its official name, was an attempt to turn over vast tracts of uninhabitable bush in Tanganyika (the mainland section of modern-day Tanzania) to sophisticated, state-of-the-art cultivation, replacing the African subsistence farmer by ranks of tractors. If the scheme had succeeded as its originators hoped, it would have satisfied a significant proportion of Britain's demand for vegetable oil, pointed the way forward for Africa to start feeding itself, and saved the hard-pressed Exchequer large amounts of money: a triple win, in other words. But it did not succeed. Instead, the scale of its failure, and the reluctance of its proponents, after so much early enthusiasm, to admit to this, turned it into a national laughing-stock for years to come. It has been written about as an ongoing, personal tragi-comedy[1] and as a warning example of foolish government meddling,[2] studied by agrarian economists,[3] employment analysts[4] and colonial historians.[5] The over-three million acres (almost 13,000 km^2, approximately the area of Yorkshire) originally planned for cultivation was whittled down year-on-year, until by 1954/55, when the Tanganyika Government took it over, the area actually cultivated stood at a mere 28,000 acres (115 km^2). The area of cultivation never exceeded its 1950 figure of 84,050 acres (340 km^2).[6] The money expended on the scheme, originally budgeted at £24 million, was most recently calculated as £46 million,[7] roughly equivalent to over ten per cent of the net expenditure on the NHS in its first year of operation, 1949/50.[8]

Figure 6:1
Proposed versus actual cultivation and expenditure on the Groundnut Scheme

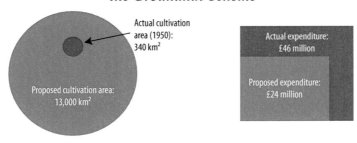

Sources: Wakefield Report (projected figures), 'The Lessons of the East African Groundnut Scheme' (actual area), *They Meant Well* (actual costs)

Individual details from the scheme are no less impressive:

- 4,000 tons of groundnuts were ordered in 1947 for seeding, but hardly any were used because of the tiny area available for planting by the end of 1948. The next year was a little better – 2,000 tons planted – but it meant that in two years the scheme had produced as crop about half of the weight it had originally bought as seed.[9]

- 'I have clear evidence that a contractor's man, buying in the Middle East before work began, not only learned from a civil engineer (who had used it) all about chain-clearing and its advantages, but actually bought 1,000 tons of precious heavy chain. He was ordered to sell his chain soon afterwards. Two years later chain-clearing was tried out at Urambo, and today regions are short of chain.'[10]

- While millions of pounds were poured into Tanganyika, the consistently good northern Nigerian groundnut crop on the other side of Africa was plagued with transport problems, particularly a lack of railway rolling stock, and significant amounts remained 'unrailed' from year to year: 92,000 tons in 1947, 154,981 tons in 1948, 137,185 tons in 1949.[11]

- Despite shortages of all kinds, there were 'enough Angostura bitters for all Tanganyika's Europeans for 70 years'.[12]

While one would hope that nothing so comprehensively comic and disastrous could befall the current low-carbon agenda, there are sufficient resonances to make the groundnut experiment extremely valuable as a point of comparison. The story is now some 60 years old, and is no longer the comedians' and cartoonists' familiar friend, a one-word joke, both feed and punchline, but its story still has much to tell us about how far a government can go astray when it believes that it is facing a crisis, and that in responding it is doing many good things at once. In short, it illustrates what can happen when everyone knows that *something must be done*, and when being right takes precedence over getting it right.

The Groundnut Scheme: an outline

Figure 6:2
The Groundnut Scheme in Tanganyika

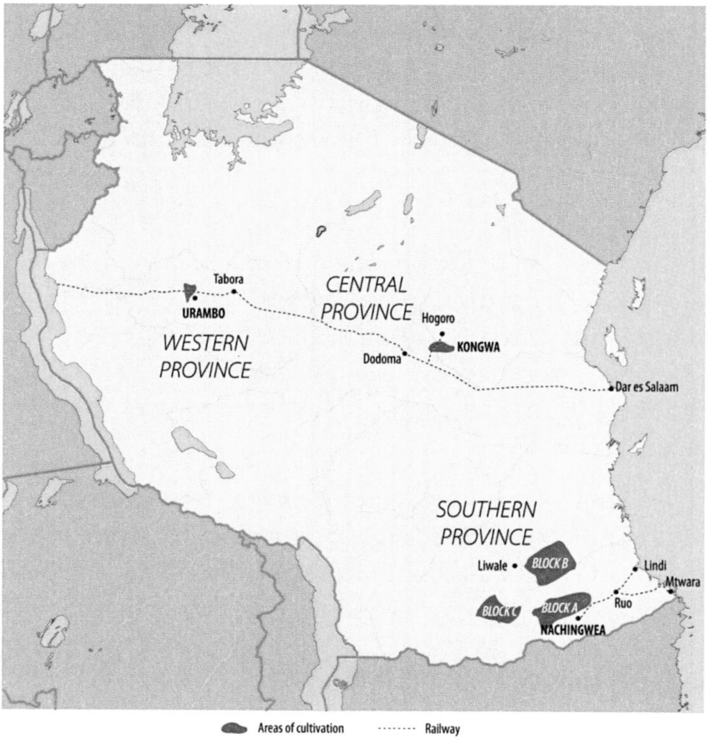

Source: Redrawn from 'The Lessons of the East African Groundnut Scheme'[13]

The outlines of the story are straightforward enough. In March 1946 the British Government, concerned that the country's supply of vegetable oil was dangerously low, received a plan from the Managing Director of the United Africa Company (UAC), Frank Samuel, for the mechanised cultivation of groundnuts in Tanganyika over an area of two-and-a-half million acres. The Government's investigations into the feasibility of Samuel's plan included a mission to Africa and the subsequent 'Wakefield Report' which increased the total area to over three million acres by recommending further farms in Kenya and Northern Rhodesia (now Zambia). By the end of the year the Minister of Food, John Strachey, had taken the decision to go ahead, and the East African Groundnut Scheme started in earnest in February 1947 when the first camp was set up at Kongwa in Tanganyika's Central Province. During its first year it was managed by the UAC while the organisation that was to run it, the Overseas Food Corporation (OFC)[14], was formed with capital from the Government of up to £50 million.

Problems began almost at once. The first batch of mostly ex-army tractors, expected in February, did not arrive until April, and then quickly started breaking down. The bush-clearing process, consisting of flattening, windrowing (to minimise soil erosion) and rooting, proved much more difficult than anticipated. In August 1947 camps were set up on the proposed sites in Tanganyika's Western and Southern Provinces (at Urambo and Nachingwea respectively). Bush clearing started in Urambo in January 1948; clearing was also begun at Nachingwea, but in the face of much internal opposition because the necessary port, railway connection and oil pipeline were all still in development. Conditions were different in each of the three areas, but the outcome was the same in all of them: desperately slow progress for much higher than expected expenses (the cost per acre in the first year worked out at around ten times the estimated amount).[15] The OFC took over from the UAC Managing Agency six months ahead of schedule, in April 1948. The miserable cultivation figures for all three areas were increased by various means, including planting in grassland as well as cleared bush in Kongwa in 1947, and subsequently supplementing the groundnut crop with sunflowers, which, while lower in oil yield, were much easier to grow, although they too fared badly.

The first six-month progress report up to 30 November 1947 surprised and dismayed MPs when they read the published version the following March. Hopes were raised by the use of new machinery in the form of the ingenious 'Shervicks', adapted from Sherman tanks, but droughts in Kongwa and Urambo ruined both the 1948 and 1949 harvests. In addition, the 1949 groundnut crop at Urambo had rosette disease (a possibility that had been mentioned when Samuel had first presented his plan). By the end of 1948, the first non-partisan murmurs of discontent could be heard, not only in the House and on Fleet Street, but amongst the administrative officers on the ground, who began to call for the resignation of the OFC Chairman, Leslie Plummer. During 1949, as the OFC continued to fail to get to grips with the scheme's various problems, these murmurs became ever louder and more distinct, and November that year proved a culminating moment. First, a full-length, illustrated and heavily publicised article on the scheme appeared in the popular magazine *Picture Post* – 'the only thorough-going account, official or otherwise, that has been given to the public'[16] – which revealed the appalling blunders made throughout the course of the scheme and called the guilty men to account. Then, only two days later, came the revelation in Parliament of the OFC's horrifying first Annual Report (covering 1 April 1948 to 31 March 1949). As a result there was a reorganisation of the OFC: the most obvious sackings were two of the three original authors of the Wakefield Report, John Rosa and John Wakefield himself. This did nothing to reduce the calls for the resignations of Strachey and Plummer.

The death-knell of the scheme had perhaps already been rung, at a dinner at the Café Royal in London on 4 January 1949, when the Chancellor Sir Stafford Cripps refused to give the scheme the extra £100 million (£2.5 billion in current values) that would have allowed it to continue on its proposed course. After this the scheme limped on, the OFC setting itself ever smaller targets, most of which still proved impossible to achieve. In 1950, the February General Election reduced the Labour Government's 1945 majority of 145 to only five and saw John Strachey replaced as Minister of Food. The rest of the year saw the OFC still desperately fumbling to renegotiate the terms of its task. By January 1951, with the publication of a new Government White Paper,[17] those terms were finally and irrevocably set out, and the scheme downgraded to the status of an agricultural 'experiment', now under the Colonial Office rather than the Ministry of Food, to run with a hugely reduced budget until 1957. Three years later, in 1954, the OFC itself was consigned to history, with the loss of 16,000 jobs, and the final three years of the repackaged scheme overseen by its replacement, the Tanganyika Agricultural Corporation (TAC).[18]

Hasty decisions

Whether it is looked at closely or far-off, the scale of the groundnut scheme disaster leads one to wonder what heights or depths of administrative idiocy could have rendered the scheme so worthless. There were, of course, plenty of mistakes made on the ground, evident in this choice paragraph from Fyfe Robertson's galvanising investigative article for *Picture Post* in late 1949:

> Enough has been written about the early blunders, most stemming from the political pressure for speed – failure to test Kongwa soils, and so discover the abrasion and tough root systems that have wrecked programmes and eaten money; 'planning' without regard to port capacity at Dar [es Salaam], or rail-and-road capacity to Kongwa; prospecting for water after choosing sites; planning tractor performance without regard to condition or spares; forgetting the need for adequate repair shops; and so on. And enough will be heard in the House about enormous stores discrepancies, complete lack of control of expenditure until last April [1948], and the failure to give information to the auditors.[19]

However, the truth was that the scheme was unworkable from the start. Alan Wood's *The Groundnut Affair* provides a much-quoted remark about the unrealised and unappreciated scale of the scheme:

> You have said the worst that can be said... and the most important thing worth saying, in pointing out that they were proposing a colossal engineering and agricultural revolution, something comparable on a small scale to the Russian Five-Year Plans, without even realising what they were doing.[20]

The speed with which the original idea was adopted implies a series of *faits accomplis*, a blithe cakewalk towards the precipice. Nevertheless, many well-informed people had given very early warning of the dangers. When Frank Samuel first presented his 'Project for the Mass Production of Groundnuts in Tropical Africa' to the then Minister of Food, Sir Ben Smith, on 27 March 1946, an agricultural expert who was present criticised Samuel's choice of location (high and dry), method of surveying (aerial) and even the groundnut itself (prone to disease, difficult to obtain). In particular, the problem with the Central Province of Tanganyika, in which half of Samuel's chosen area lay, was rainfall, or rather the lack of it. Samuel disputed this, beginning a recurring pattern of claim and denial that would muddy the rainfall question throughout the investigative period.

In any case, despite this opposition, Wakefield's three-man mission was on its way to Tanganyika by June. Even at this early stage, Frank Sykes, an expert on mechanised agriculture, had become a 'Doubting Thomas about the East African scheme' on first learning of it, thinking it:

> unjustifiably optimistic in the estimate of yields, in the low number of tractors, etc., employed, and in the belief that weeding can be avoided. In fact, he has said... that the groundnut is not really a good crop for mechanisation at all.[21]

Sykes would prove to be correct on every count.

Contemporary commentators would later find ways to exonerate the mission members from too much blame. It is certainly true that they were given precious little time to complete their work: the Wakefield Report was presented to the Colonial Minister on 20 September 1946, exactly three months after the mission had left England, and their nine-week schedule forced them to view most of the area under investigation from the air (10,000 miles, compared with 2,000 miles by road and 1,000 by rail). However, the evidence suggests that they went out of their way to look for the good news as they sent their optimistic cables back to London. For instance, the area that would finally be chosen for the start of the scheme, around Kongwa in the Central Province, was an afterthought for the mission (perhaps the experts' warnings about the low rainfall had been listened to on this occasion) but was added to the Wakefield Report's list of areas to cultivate through the agency of one man, Tom Bain. Wood even calls Bain's introductory letter to Wakefield – 'probably the most important ever written in the whole history of the groundnut scheme':[22] in it, Bain invited the mission to come and see the groundnuts on his farm near Kongwa. The report goes so far as to mention the weight of this crop, '1,200 to 2,000 lb of shelled nuts per acre', twice.[23] The 'marginal' levels of rainfall in Kongwa, a region known to local natives (who were presumably not asked) as 'the country of perpetual drought',[24] were even mentioned in the Report, but its potentially bad effects were outweighed by the evidence of Tom Bain's groundnut crop:

> On paper, at least, the rainfall of the Northern Mpwapwa area [Kongwa] appears to be marginal, but the results given above speak for themselves.[25]

The mission (the non-Government member of which was the UAC Plantations Manager, David Martin) also sought advice from A.L. Gladwell, a man who had gained extensive bush-clearing experience during the war, and who, as the Managing Director of UAC subsidiary Gailey and Roberts, had been asked to extend them all the help he could. After discussion with Gladwell, the Report set the cost of clearing the bush at £3 17s 4d per acre – perhaps the most misleading statistic in a document whose wrong-headed calculations would be repeatedly exposed.

However, when the mission met with officials to discuss their findings on their return to England in early September, the only part of their remarks with which anyone explicitly took issue was their estimate for average yields: Wakefield's suggestion of 850 lbs per acre seemed rather high to the experts, and would eventually be reduced to 750 lbs. Technical questions remained to be answered, it is true, but the meeting nonetheless agreed to go ahead with the scheme as it stood and to begin it as soon as Cabinet gave their approval.

Up to this point the scheme had looked very much like a Colonial Office project: both Wakefield and the third member of the mission, the financial expert John Rosa, had come from that Department. However, at the suggestion of the Colonial Secretary Arthur Creech Jones,[26] the scheme was passed onto the Ministry of Food, now under John Strachey, who initially required a thorough investigation into the scheme, and thought that no more than £3 million should be committed to it until that investigation was finished. However, Creech Jones, after speaking to Frank Samuel, was able to persuade him that this would prevent the scheme from starting the following year. He would soon be an enthusiastic and unquestioning supporter, partly because of his sense that the world and Britain in particular was facing a food crisis: David Low referred to him as 'Starvation Strachey' in an *Evening Standard* cartoon of 10 November 1949.[27] Strachey's conversion was typical, and while there were still dissenting voices – Herbert Morrison's economic adviser asked whether it was 'really necessary to buy this £3 million pig-in-a-poke'[28] – a large number of Ministers rapidly became 'besotted',[29] the Chancellor Hugh Dalton being so keen on the scheme that he happily ignored the objections of some of his advisers.

As a sop to Strachey's and others' concerns, a Special Section of the Ministry of Food (including two familiar faces, Wakefield and Rosa) was formed to look into the report's viability:

> On 31 October, His Majesty's Government, recognising the urgency of taking every possible step to secure an increased supply of fats, decided that the Ministry of Food should put in hand immediately all the necessary preparations for carrying out the first year's work proposed in the Mission's Report, and should concurrently undertake a detailed investigation of the long-term plan in all its aspects, particularly the financial aspects, the availability of equipment, the availability of agents, staff and labour, the problem of communications, the attitude of the local Governments, and other matters requiring detailed examination before a decision on the full scope of the scheme could properly be taken.[30]

The 'detailed investigation' engaged in by the Special Section was completed swiftly, as Strachey announced the scheme to the House of Commons within a month, and the White Paper (Cmd 7030) was published on 5 February 1947, the day after the scheme's advance party arrived in Tanganyika. The fact that it was done 'concurrently' with making preparations to put the scheme into action foreshadows the disastrous dovetailing of research with practice that was to happen on the ground in Tanganyika.

The Special Section's report (in effect, the report on the [Wakefield] report on the [Samuel] report) contained some warning voices, such as those of the chemist W. M. Crowther and agriculturalist Dunstan Skilbeck, who questioned the figures for rainfall, but the general tone was upbeat. In particular, the conclusions, credited to F. Hollins, showed great confidence in the financial soundness of the scheme, which events would quickly prove to be very mistaken:

> Viewed strictly as a commercial proposition, the scheme involves no unjustifiable finance risks.[31]

The scheme as a military operation

As if it wasn't enough that the premises on which it had been set up were so shaky, the groundnut scheme seems to have been based on a number of apparent contradictions. First, there was an unmistakable but inappropriate military aspect to the scheme, a khaki line that implicitly and explicitly ran through it. Historian John Iliffe has remarked that 'the operation acquired the ex-army-surplus quality which pervaded post-war Britain'.[32] As Hogendorn puts it:

> The project… assumed something of a military character. Its field director was a general [Desmond Harrison, Chief Engineer in the Burma Campaign under Mountbatten], and many of its black and white workers were fresh from military service. The army was even represented in the choice of tractors, with the famous converted tanks, 'shervicks', constituting a fair percentage of the force.[33]

According to his opposite number Alan Lennox-Boyd, John Strachey 'invariably talked of the groundnut scheme in army metaphors, using military illustrations'.[34] Strachey himself introduced the scheme to the House in December 1946 with an explicit comparison:

> In many respects, the planning and carrying out of this scheme resemble a military operation. We are working against time and we have to move very large quantities of constructional machinery and equipment, mobile workshops, heavy tractors, etc., over difficult country, while maintaining all the personnel and equipment ourselves, in much the same way that a military formation has to supply all its own maintenance services.[35]

The following November he would refer to the morale-boosting speech he had made at Kongwa:

> On your success depends more than on any other single factor whether the harassed housewives of Great Britain get more margarine, cooking fats and soap in the reasonably near future. I believe that the United Africa Company and all those who have been officially and unofficially responsible for the very rapid launching of this scheme of 'Operation Groundnuts', as I have called it, deserve well of the people of Great Britain. One inevitably uses a military term such as operation in describing this scheme, because almost the only analogy of an operation on this scale is provided by the military sphere. The East African Groundnuts Operation is a great expedition, and I can never help comparing and contrasting it with the other great expedition in North Africa, the landings in 1942.[36]

John Wakefield, later described as one of the 'amateur strategists'[37] behind the scheme, talked about its resemblance to the Mulberry harbours operation before D-Day. Behind all this is the idea of fighting for a great cause, against an enemy – the implacable African landscape

– that will give you no quarter. But perhaps this was just another aspect of the scheme's seductive properties, and the sheer difficulty of the task another reason to believe that it was the right thing to do.

Of course, as Alan Wood, himself no stranger to comparing the scheme to a military operation, commented:

> The Groundnut Army was in fact, in very much the same position as an army which had gone into action in a fit of absent-mindedness, and forgotten to take any RASC with it.[38]

Invoking the military aspect was interpreted by those opposed to the scheme as another way of saying 'Nuts at all costs',[39] an argument which Charles Ponsonby rejected:

> This scheme is always referred to as a military operation, and military operations are obviously wasteful affairs, but on this occasion the money is the taxpayers' money, which we are voting today, and expenditure should be watched exactly as if it were a great private corporation.[40]

The scheme as an experiment

When the scheme wasn't being compared to a military operation, it was being referred to as an experiment. Strachey, at first uncertain, was able to go along with the lack of thoroughness of the Special Section's investigation into the scheme by assuring himself that it was, at first, a 'large-scale experiment', a term that appears in the White Paper, rather than a going concern. In other words, the scheme was its own pilot scheme. This allowed him to be consistent when the first year proved a disaster: from 1948 the scheme could start anew, especially with its authentic administrator, the Overseas Food Corporation, in place from 1 April. It might also explain Leslie Plummer's remark that 'the way in which you discover snags attendant on large-scale mechanised agriculture is to proceed with large-scale mechanised agriculture'.[41] (Insouciance of this variety lives on in hopes for 'learning by doing' prevalent in so much of the modern environmentalist literature.)

Who was in charge?

Another, less theoretical, area of confusion was the placing of the project under the authority of the Ministry of Food rather than the

Colonial Office. This surprised even sympathetic MPs, and there would be a number of clashes in Tanganyika between the Government there, supported by the Colonial Office, and the Overseas Food Corporation, supported by the Ministry of Food. The latter had been put in charge of the project because many believed that its key aims was the supply of oils to Britain, and that the improvement in African agriculture was no more than a by-product. Others believed that they were equally important aims. Eventually, when the scheme had patently failed to produce the promised oil supply, it had no choice but to become all about African agriculture. There were also clear divisions between those who believed that the oil shortage was acute and short-term, and those who thought that it was only going to get worse with the increasing pressure from a growing population.

There was also a conflict contained in Frank Samuel's original plan, which hoped that the project would bring economic benefits to the country without disturbing the way of life of Tanganyika's population. This must have been partly related to the fact that Tanganyika was a UN Trust Territory rather than a colony, and would also have chimed with a socialist administration uncomfortable about being accused of exploitation.[42] Certainly, a great deal of care was taken not to interfere with native rights: one reason the scheme's advance party chose Kongwa over nearby Sagara, with its plentiful water supply, was that the latter was used by the local Wagogo tribe for watering their cattle. However, Samuel's sentiments must have seemed impossibly utopian when Kongwa became a sort of 'boom town': 'mushroom villages of an insanitary type, which were the haunts of prostitutes and other undesirables, had sprung up in the vicinity of the groundnut camp'.[43] For a while it had the largest population of any centre in Tanganyika outside Dar es Salaam, accruing the usual disadvantages of such a condition: high crime (its newly built police station was immediately swamped), high prices for staple goods and a huge influx of uninvited people. There seems no question that the capital poured into the groundnut project had an adverse effect on other parts of the Tanganyikan economy, in particular on Tanganyika's main crop, sisal,[44] a huge employer and worth about £3,800,000 in 1947. Charles Ponsonby was particularly troubled about rumours that 'the Ministry of Food are claiming very high priority and snatching everything in the way of machinery and materials that they can lay their hands on',[45] while his fellow MP J.H. Hare complained that the sisal industry was suffering 'directly from shortage of manpower and machinery owing to the priorities given to the groundnuts scheme'.[46]

Doing the right thing

Given the problems that emerged so early in the scheme's life, it may seem surprising that political and public support continued for so long, often in the face of clear contrary evidence. Incompetence and vanity doubtless played their part, but, on balance, it seems that most were simply misled by the variety and depth of the benefits apparently on offer. The public proved as enthusiastic as anyone: 100,000 men applied to join 'the groundnut army', many of them only recently demobbed. In 1949 the MP Harry Crookshank, never one of the believers, recalled the misplaced enthusiasm of the early days:

> Never, I should think, in the history of any Government project was there so much advertisement and propaganda and general anticipatory declaration as there have been with regard to this scheme. Indeed, ever since it was first announced there has been hung over it a haze of rejoicing, as if everything were now finished, as if we were getting the product and as if the plan had already been carried out.[47]

The points of comparison with contemporary government proposals for low-carbon energy generation are numerous. As a well meaning, multiple win, the scheme rose above criticism. The fact that it *deserved* to succeed all but completely obscured the likelihood of failure. Groundnuts promised to help Britain's 'harassed housewives' by combating world food shortages, deliver a philanthropic improvement in African agriculture, create employment for British citizens, and bring Britain much needed financial relief via the relative cheapness of the East African crop. This latter point was critical. In Parliament Dr Edith Summerskill, on behalf of the Ministry of Food, did the calculations for her fellow members using the Wakefield Report projections:

> The first harvest will be in 1948, and the crop is estimated to be 50,000 tons. In 1950 to 1951 we hope to get 600,000 tons, and later 800,000 tons. So far as the cost of production is concerned, it will cost £14 5s. 6d. per ton, while today's price is £32; and this margin of approximately £17, applied to 600,000 tons, will mean a saving to this country of £10 million.[48]

In the straitened circumstances of the late 1940s this was an enormous sum, but as A.T.P. Seabrook, who was one of the OFC managers on the ground in Tanganyika (and would later be the Chief Administrative Officer of the TAC), was to observe, 'The economics looked too good to be true, as indeed they were.'[49]

More important for the continued support of the scheme was surely the philanthropic desire to put Africa on the road to self-sufficiency, as laid out in the 1947 White Paper:

> While the immediate reasons for the launching of this scheme is the urgent need for new supplies of fats for the United Kingdom, His Majesty's Government believe that its long-term importance may lie even more in the practical demonstration it will provide of the improved productivity, health, social welfare and prosperity which scientific agriculture can bring to Africa.[50]

A small but telling example of the reach of this high-minded idea is that the scheme's Agricultural General Manager from April 1948, the South African John Phillips, was urged to take the post by his country's President, Jan Smuts, 'on the grounds that the task ahead was one of great importance for all Africa'.[51] Even Alan Wood himself, despite his profound misgivings and personal reasons for disliking the scheme, continued to support the principles behind it. His brief preface to *The Groundnut Affair* is perhaps worth quoting in full:

> This, in large measure, is a story of failure, frustration, heartbreak, bad luck and bad blunders. It tells of a tragedy, with many of the elements of a tragi-comedy. But the story starts as one of the most inspiring ventures since the Second World War: and it may yet prove to be one of the most worthwhile experiments now being undertaken in a mad world already talking of more wars to come.[52]

Seabrook saw the optimism and enthusiasm for the scheme as 'an almost fanatical faith in ultimate success', adding, 'the setbacks experienced were only the teething troubles to be expected'. He offered a pointed rationalisation, however:

> I readily admit that I shared that faith to the full, largely I think because the difficulties were not – or did not seem at that time to be – insuperable, and also because of a rather simple faith that something that had been worked out by a group of experts, subjected to further expert examination, unanimously applauded by all political parties in the British House of Commons and by the public... just could not be misconceived and therefore impracticable.[53]

While the difficulties did indeed prove intractable and the tide turned against the scheme, nobody seems to have had a bad word to say about the ordinary 'groundnutters' and their heroic enterprise, and certainly Fyfe Robertson's admiration of the 'groundnut army' is unequivocal:

> Out there in Tanganyika an army of young men have been fighting a battle the like of which has never been known before… Here is a story of attack and counter-attack, of heartbreak and retreat, and of stubborn, slow, but quickening successes.[54]

Robertson compares the scheme to Dunkirk – a defeat turned into something more positive, rescued by a thousand acts of individual bravery or ingenuity. A better military comparison might have been Thermopylae: a small group of heroes, fighting on, but inevitably succumbing to insuperable odds, and ultimately achieving very little. A more prosaic way of looking at the scheme was entertainingly provided by Charles Ponsonby:

> [The scheme] is also a great speculation. Many of us who are going to put our money on a horse take an interest in the name and pedigree of the horse. In this case, the horse is 'Speculation' by 'Government' out of 'Necessity'.[55]

It was the excitement, as well as the desperation, of this speculation that seduced Frank Samuel into expanding its parameters by a scale of 25, an extraordinary folly given that the plan he had originally received from R.W.R. Miller, Director of Agriculture in Tanganyika, had suggested a not unambitious area of 100,000 acres. It allowed John Wakefield, during his mission, to make the best of the situation on the basis of individual examples and anecdotes. In the case of John Strachey, who was the driving force of the scheme in and out of Westminster, his attachment to the idea, after his early caution, warped into a stubborn, occasionally dishonest obliviousness when the scheme began to fall apart.

Conclusion

Reviewing the project with the low-carbon economy in mind, we can see a number of important points of comparison. Those proposing the groundnut scheme believed that they were facing a crisis, and that emergency action driven by the state was the only possible response.

This frame of mind readily tended towards a misconception of its project as being military, leading to a situation where cost was of no concern. In fact, the groundnut scheme was nothing if not commercial, but cheaper and manifestly more practical alternatives such as the West African potential were simply ignored because of the promise of larger prizes in Tanganyika, and, in their haste and zeal, officials paid scant attention to efficient management of costs during the development.

Perhaps most interestingly of all, reports on the feasibility of the scheme were always inclined to accentuate the positive, and those expressing concerns, and there were many, were marginalised, as if criticism were in some sense disloyal or a breach of faith. At the technical level there was a tendency to overstate the importance of theoretical maximum clearance and planting rates, leading to infeasible targets, which then resulted in erroneous estimates of yields and income. Too little attention was paid to cheaper but less dramatic means of achieving the same ends, such as international trade or improving transport in West Africa to move an existing peanut crop.

Ultimately, it must be remembered that the Groundnut Scheme failed not only in the ledger books of its accountants, but still more acutely in the bush of Tanganyika amongst a litter of broken tractors and drought stricken fields. And it failed because theoretical investigation of the conditions to be faced was optimistic and mistaken, and because the investigators presumed that the necessary technologies already existed, and where deficiencies remained they could simply 'learn by doing' on the grandest scale. None of these assumptions was correct.

PART THREE

Two Case Studies

Chapter Seven
Germany's Cloudy Future

Introduction

Germany's longstanding and widespread commitment to environmentalism is evident both in its pioneering recycling programmes,
including its 'dual system' for waste collection and its *Grüne Punkt*
certificate for recycling packaging, and also the visibility and
influence of its environmentalist party, Bündnis 90/Die Grünen
(Alliance 90/The Greens), which has more than ten per cent of the
seats in the Bundestag (German Parliament). In its support of
renewable energy sources, a support that, as Europe's wealthiest
country, it has perhaps been better able to afford than its neighbours,
it is truly a leader. Germany has the world's largest installed capacity
of photovoltaic cells, and its wind capacity is greater than that of any
other country except the United States:

Figure 7:1
Installed capacities of wind power and PV in Germany,
USA and Spain, 2008

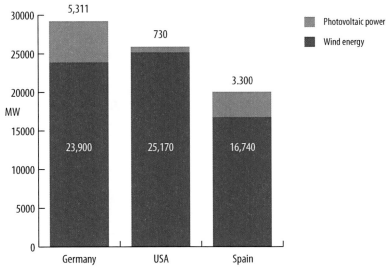

Source: 'Economic Impacts from the Promotion of Renewable Energies' (2010)[1]

Germany's renewables policy has maintained the impression of a success story, and, more to the point, a replicable one, whereas other states have experienced clear difficulties, such as Spain, or appear to be one-off special cases such as Brazil or Sweden. For a long time Germany has appeared to be the shining example of how to develop a renewable energy sector, and has been watched with interest and admiration by commentators in the UK and US.[2] However, there has long been a slight undertone of bafflement about this success, as can be inferred from the title of one positive *Washington Post* article from 2007, 'Cloudy Germany a Powerhouse in Solar Energy', which suggests, perhaps inadvertently, that this success is against the odds and to be regarded with circumspection.

If the shine has begun to come off in the past twelve months, the German state remains ebullient, even triumphant in its official publications, for example those of the Federal Ministry for the Environment, Nature Conservation and Nuclear Safety (Bundes-ministerium für Umwelt, Naturschutz und Reaktorsicherheit – BMU).[3] Germany's first offshore wind farm, Alpha Ventus, buoyed up by newly increased offshore subsidies (the feed-in tariff was raised from 9 to 15 cents/kWh in 2009: see Table 7:1, p. 77), began operation in the North Sea on 27 April 2010. Many of Germany's renewables targets for 2010 have been met: in particular, the target for electricity generation (12.5 per cent of MWhs by 2010) was achieved as early as 2007, and by 2009 the renewables share was as much as 16.1 per cent:

Figure 7:2
Renewable energy share of electricity generation, 2009

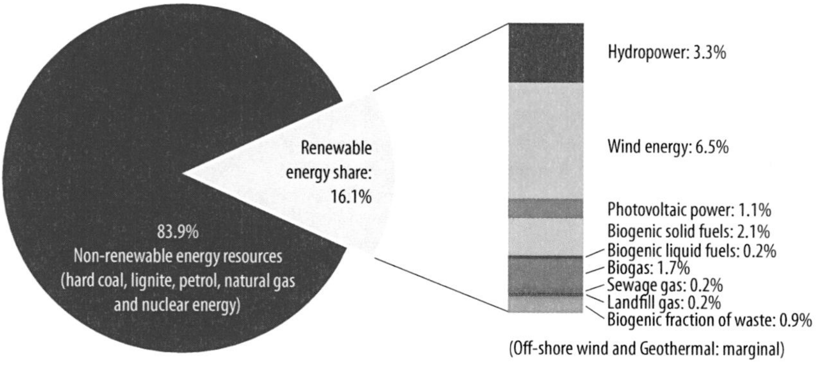

Hydropower: 3.3%

Wind energy: 6.5%

Photovoltaic power: 1.1%
Biogenic solid fuels: 2.1%
Biogenic liquid fuels: 0.2%
Biogas: 1.7%
Sewage gas: 0.2%
Landfill gas: 0.2%
Biogenic fraction of waste: 0.9%

(Off-shore wind and Geothermal: marginal)

Renewable energy share: 16.1%

83.9%
Non-renewable energy resources
(hard coal, lignite, petrol, natural gas
and nuclear energy)

Figures from *Renewable Energy Sources in Figures*[4]

The rise in the renewable share of the energy mix has meant that the BMU is able to make other impressive claims: for instance, that greenhouse gas emissions have been reduced, by no less than 107 million tonnes of CO_2 in 2009,[5] and that an estimated 300,000 people were employed in the renewable energy sector by 2009:

Figure 7:3
Employees in Germany's renewable energies sector, 2004, 2008, 2009

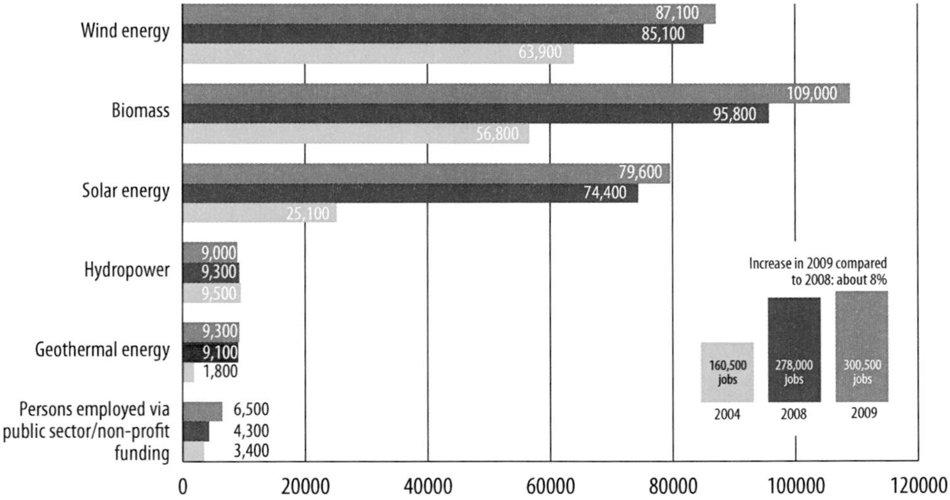

Source: *Renewable Energy Sources in Figures*[6]

Feed-in tariffs: a success?

Germany can even be considered to have set the European agenda in renewables policy, with the subsidy system that it has used for two decades, feed-in tariffs, being adopted by a number of European countries, including France, Italy, Spain and, most recently, the UK, though the new Government already seems to be changing its mind, and on 7 February 2011 the Energy and Climate Change Minister Chris Huhne announced an urgent review into a scheme which is not yet a year old. Indeed, this review announced its findings just over a month later, and proposed revisions, particularly to the tariffs for larger installations of solar photovoltaics, that would reduce the cost of the scheme to consumers by between £2bn and £4bn over a ten-year period.[7]

Feed-in tariffs work by obliging electric supply companies to pay a premium to generators of renewable electricity, with the utility companies then passing this increased cost onto their customers: out of an average German household electricity price of 23.2 euro cents per kWh, the feed-in tariff accounts for 1.2 euro cents.

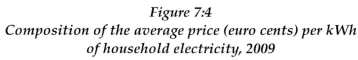

Figure 7:4
Composition of the average price (euro cents) per kWh
of household electricity, 2009

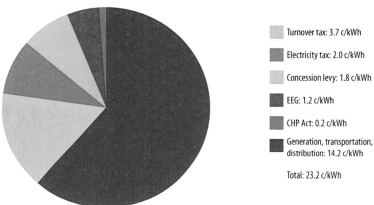

Turnover tax: 3.7 c/kWh

Electricity tax: 2.0 c/kWh

Concession levy: 1.8 c/kWh

EEG: 1.2 c/kWh

CHP Act: 0.2 c/kWh

Generation, transportation, distribution: 14.2 c/kWh

Total: 23.2 c/kWh

Source: *Renewable Energy Sources in Figures.*[8] (The EEG is the Renewable Energy Sources Act [Erneuerbare-Energien-Gesetz], the legal basis for the feed-in tariffs)

The German tariffs vary according to the renewable energy source – wind, photovoltaic, biomass, and so on – and can be paid to domestic providers, such as PV panels on roofs, as well as industrial-sized power plants. A key principle of the feed-in tariffs is degression: in other words, as the take-up for the renewable technologies increases, the set feed-in tariffs for newly installed plant are expected to be reduced, typically by five per cent a year. However, Table 7:1 shows how the general trend of these reductions has been interrupted by fresh tariff figures for some technologies in 2004 and 2009.

Table 7:1
Technology-specific feed-in tariffs in euro cents/kWh

	2000	2001	2002	2003	2004	2005	2006	2007	2008	2009
Wind (onshore)	9.10	9.10	9.00	8.90	8.70	8.53	8.36	8.19	8.03	9.20
Wind (offshore)	9.10	9.10	9.00	8.90	9.10	9.10	9.10	9.10	8.92	15.00
Photovoltaics	50.62	50.62	48.09	45.69	50.58	54.53	51.80	49.21	46.75	43.01
Biomass	10.23	10.23	10.13	10.03	14.00	13.77	13.54	13.32	13.10	14.70
Mean Tariff	8.50	8.69	8.91	9.16	9.29	10.00	10.88	11.36	12.25	13.60

Source: 'Economic Impacts from the Promotion of Renewable Energies' (2010)[9]

The tariffs have been refined over time by political directives, beginning with the Electricity Feed-in Law (Stromeinspeisegesetz – StrEG) in 1991, which created the feed-in tariff system. This was amended by changes to the Federal Building Code (Baugesetzbuch – BauGB) in 1997, and then replaced by the Renewable Energy Sources Act (Erneuerbare-Energien-Gesetz – EEG) in 2000. The EEG, which set the familiar targets of 12.5 per cent renewable electricity generation by 2010, and 20 per cent by 2020, was itself amended in 2004 and once again in 2009. A further revision to the EEG is expected in 2012. The effect of the various directives on the growth of electricity generation from renewables is shown in Figure 7:5.

Although the installed capacity (MWs) of renewable energy has also increased significantly in the last decade (Figure 7:6), electrical energy generation (MWhs) is a more informative quantity, since the load factor of renewables (i.e., the proportion of actual output to theoretical maximum output over a period) tends to be low. The annual load factors of Germany's PV power stations have never exceeded 13 per cent, with most of them stubbornly stuck at around 10 per cent. Even the BMU's upbeat *Renewable Energy Sources in Figures* has to admit that 2009, because it had 'unusually light wind conditions', showed poor results for electricity generation from wind, with a year initial capacity of 23,860 MW producing only 37.8 TWh, implying a load factor of less than 18 per cent.[11]

Figure 7:5

Development of electricity generation from renewable energies since 1990

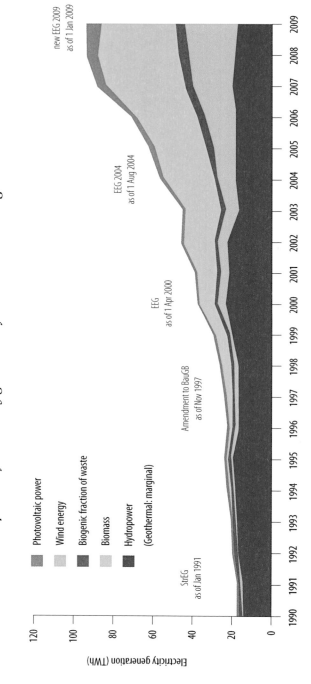

Source: *Renewable Energy Sources in Figures*[10]

Figure 7:6
Shares of total renewables-based installed capacity in the electricity sector, 2000 and 2009

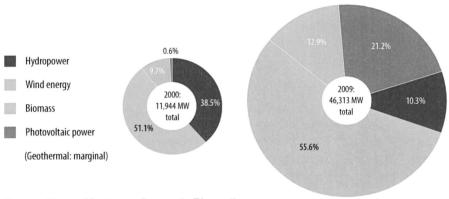

Source: *Renewable Energy Sources in Figures*[12]

Critics of the FIT system

In spite of the apparent success of the feed-in tariffs, schemes of this kind have aroused increasing criticism in other countries, with one of the most vocal detractors in Britain being the leading green pundit George Monbiot,[13] and they have not gone unchallenged in Germany, where leading economists such as Wolfgang Pfaffenberger of the Bremer Energie Institut at the University of Bremen have commented on the likely distinction between the gross economic impacts, in terms of GDP and employment, of renewable subsidies, and the net impacts resulting from the suppressive effects of increased energy costs.[14]

Further work by the German economist Manuel Frondel, based at Ruhr University but also head of the Rheinisch-Westfällisches Institute for Economic Research (Rheinisch-Westfällisches Institut für Wirtschaftsforschung, RWI), has built on this growing literature, and can now be conveniently read in a 2010 paper in *Energy Policy*, 'Economic impacts from the promotion of renewable energies: The German experience', jointly authored with colleagues from Jacobs University, Bremen, and the RWI.[15] The summary conclusion of the paper is simple:

> We argue that German renewable energy policy, and in particular the adopted feed-in tariff scheme, has failed to harness the market incentives needed to ensure a viable and cost-effective introduction of renewable energies into the country's

energy portfolio. To the contrary, the government's support mechanisms have in many respects subverted these incentives, resulting in massive expenditures that show little long-term promise for stimulating the economy, protecting the environment, or increasing energy security.[16]

This work originally appeared as an RWI working paper,[17] with minor textual differences, including references to US-specific studies, and one or two extra charts, but also with an executive summary and a comprehensive appendix of tables. In this form the report elicited a number of reactions, including a swift governmental rejoinder on the BMU website.[18] This was brief and, even within its brevity, inadequate, trying to deflect the thrust of the RWI paper's arguments with a question-begging claim that: 'It just brings up well-known arguments against the successful EEG that have been refuted a long time ago.' The paper was intensely controversial in Germany, where it seems to have been considered unpatriotic, and an interviewer from the TV news programme *Monitor* on the German WDR channel suggested that the study was compromised since the work had been funded by the Institute for Energy Research (IER), based in Texas,[19] which has links with the conventional energy industry. However, as has already been noted, the work of Frondel and his colleagues has roots in a literature that predates the IER funding, and in any case has to be judged on its own merits.

Frondel and his co-writers launch their critique on a number of related fronts:

- The high year-on-year cost of the subsidies, especially in comparison with the amounts given to research and development in the renewable energy sector

- The current and future burden passed on to electricity consumers

- No impact on overall European CO_2 emissions

- The negligible (even negative) impact on German employment

- The stifling of technological innovation in the renewables sector

Costs

The sheer expense of the feed-in tariffs seems unarguable. The total over the eight years between 2001 and 2008 amounts to €36.65 billion.

Table 7:2
Share of feed-in tariff expenditures allocated to major technologies, 2001–8

	2001	2002	2003	2004	2005	2006	2007	2008
Wind Power	–	64.5%	64.1%	63.7%	54.3%	47.1%	44.5%	39.5%
Biomass	–	10.4%	12.5%	14.1%	17.7%	23.0%	27.4%	29.9%
Photovoltaics	–	3.7%	5.9%	7.8%	15.1%	20.3%	20.2%	24.6%
Total, Bn €	1.58	2.23	2.61	3.61	4.40	5.61	7.59	9.02

Source: 'Economic Impacts from the Promotion of Renewable Energies' (2010)[20]

The figure for 2007, 7.59 billion euros, can be directly compared with the same year's 211.1 million euros of government investment and 138.5 million euros of private investment in renewable energy sector research and development.[21] In other words, R&D funding in the sector was less than five per cent of the expenditure on feed-in tariff income support for existing renewable generators.

As Table 4 above shows, the PV tariff has always been an order of magnitude higher than the others, despite, or rather (since the feed-in tariff system is designed to favour technologies that appear to need a 'helping hand') because of, the relatively small contribution of about one per cent it makes to the electricity mix. The high PV tariff is one of Frondel's main targets; indeed, an earlier RWI paper, 'Germany's Solar Cell Promotion: Dark Clouds on the Horizon',[22] focuses on photovoltaics exclusively, suggesting a comparison, if PV tariffs were to continue at such a high rate until 2020, with the disastrous long-term subsidising of German hard coal. It can be pointed out that the hard coal subsidies were narrowly focused on preserving an industry, and, though this is not to excuse them, were at the expense of taxpayers not consumers. Indeed, generally speaking, subsidies to the energy sector, now under concerted criticism from institutions such as the International Energy Agency,[23] are of this latter kind, namely consumer-subsidies holding prices at artificially low levels (for example, gasoline in Venezuela) by transferring the burden, usually to taxpayers, with a loosely progressive motivation. As the International Energy Agency (IEA) remarks: 'Energy subsidies... are often used to alleviate energy poverty and promote economic development by enabling access to affordable modern energy services.'[24] Subsidies to energy producers at the expense of energy consumers, as observed in the renewables, are not only different in character and macro-economic impact, but are also rather more politically sensitive, a point to which we will return when considering 'willingness to pay'.

The tendency for these consumer impacts to be backloaded, and cumulatively significant, rather than instantly acute, is an important consideration. The RWI study observes that when the EEG was enacted in 2000, it guaranteed 20 years of subsequent tariffs at the rate of the tariff when the producer began generating electricity. Clearly intended as an incentive for 'early adopters', it means that a producer who began generating energy from PV in 2000 can expect to receive a tariff of 50.62 euro cents per kWh (i.e. the 2000 level of tariff) until 2020, even if the 2020 feed-in tariff itself is down to single figures. Frondel and his colleagues map out the continuing payments to the end of 2029 in a particularly startling chart:

Further, Frondel calculates that, by 2010, the increase in PV installations during the previous decade will entail a burden of around 52 billion euros which, with the current cost for new installations of 13 billion euros, will mean a total burden of 65 billion euros.[26] These calculations are complex and potentially controversial, taking account, for instance, of the likely increase in electricity prices over the succeeding years, which will bring the actual price ever closer to the guaranteed feed-in tariff and, in the case of lower tariff technologies such as wind, may even surpass it.

Emissions

Examining the cost of CO_2 emissions abatement by means of PV, Frondel further calculates that the costs could be as high as 716 euros per tonne of CO_2, which compares unfavourably with the price of certificates in the European Union Emissions Trading System (EU ETS), 30 euros per tonne. Moreover, the very existence of the ETS cap and trade system will logically negate any extra CO_2 savings the EEG could hope to make: as Frondel puts it, the effect 'is merely a shift, rather than a reduction in the volume of emissions'.[27] A recent study

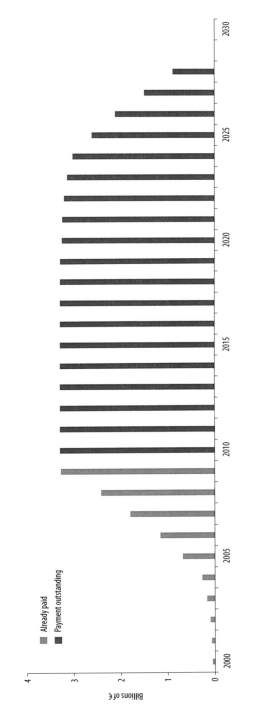

Figure 7:7

Annual amount of feed-in tariffs for PV modules installed between 2000 and 2009

Source: 'Economic Impacts from the Promotion of Renewable Energies' (2010)[25] by Traber and Kemfert appears to support this theoretical observation, showing that, while German emissions have indeed been reduced, overall European emissions have stayed much the same.[28] Certainly, amongst economists it is not controversial that the renewables policy has no additional emissions-reducing effect over and above the EU ETS, which caps such reductions. As Professor David Newbery at Cambridge's Energy Policy Research Group puts it: the 'current ETS sets the quota of total EU emissions... Increased R[enewable] E[nergy] S[upply] does not reduce CO_2'.[29]

Employment

The question of employment in the renewables sector, which is a key element of the 'win-win-win' scenario put forward by Mr Cameron, is particularly complicated. That there are new jobs, as shown in Figure 7:3 (p. 75), is not in question, nor is it in question that, like most increases in employment, they will produce further ancillary employment, at least in the short run. What is open to doubt is whether these jobs are offset by the loss of jobs in other areas, and by how much. It is possible to theorise where such losses, which are far from confined to the German situation, would come:

- Balancing job losses in the conventional energy sector as less of that sector is used

- Job losses, through cut-backs and business closures, in industry and other heavy energy users caused by the renewable energy-related increase in energy costs

- Job losses, through cut-backs and business closures, in the wider economy caused by the knock-on effect of higher energy costs on consumer spending

- Job losses due to a reduction in the businesses needing high capital investment, as that investment is diverted to the subsidised renewable energy sector

Frondel and his colleagues note the Ministry's claims that renewables were a 'job motor for Germany', and the large number of employees attributed to the sector, 249,000, with the expectation that this will rise to 400,000 in 2020, but write:

While such projections convey seemingly impressive prospects for gross employment growth, they obscure the broader implications for economic welfare by omitting any accounting of off-setting impacts. The most immediate of these impacts are job losses that result from the crowding out of cheaper forms of conventional energy generation, along with indirect impacts on upstream industries. Additional job losses will arise from the drain on economic activity precipitated by higher electricity prices. In this regard, even though the majority of the German population embraces renewable energy technologies, two important aspects must be taken into account. First, the private consumers' overall loss of purchasing power due to higher electricity prices adds up to billions of Euros. Second, with the exception of the preferentially treated energy-intensive firms, the total investments of industrial energy consumers may be substantially lower. Hence, by constraining the budgets of private and industrial consumers, increased prices ultimately divert funds from alternative, possibly more beneficial, investments. The resulting loss in purchasing power and investment capital causes negative employment effects in other sectors, casting doubt on whether the EEG's net employment effects are positive at all.[30]

These views are consistent with earlier complex economic modelling by Hillebrand dating from 2005,[31] which suggested that an initial increase in jobs owing to the expansion of renewable energy (the 'investment effect') would begin by increasing employment overall (plus 33,000); then, due to the raised cost of electricity (the 'cost effect'), the employment balance would even out, until, by 2010, it would be slightly negative (minus 6,000), a point about which there seems to be general agreement in a number of other contemporary economic studies.[32] Hillebrand's observations on the timing of these effects is important, and should be borne in mind when considering the political sensitivity of support mechanisms for renewables:

> On balance, while the investment-induced demand effect dominates in the first years, the contractive cost impulse will prevail thereafter.[33]

As Kipling remarked in one of his most vatic warnings, 'Morning never tries you till the afternoon'.[34]

Innovations and the fall in solar prices

The charge that the EEG has stifled technological innovation in the renewables sector is, by comparison with the employment question, straightforward to investigate. The fact that, as previously mentioned, a generator is entitled for 20 years to the level of feed-in tariff at which they began, and that there was an expected (though, in the end, unpredictable) annual degression of five per cent in the tariff, suggests that the producers would have rushed to construct as soon as possible and thus receive the highest level. This is hardly a situation conducive to a considered plan, involving a period of R&D: indeed, it seems likely that producers would have been tempted to choose inferior or even incomplete technologies in order to take advantage of the subsidy structure. Stuart Wenham, Chief Technology Officer at the Chinese solar cell producer Suntech Power Holdings, remarked:

> The industry was getting into a situation two years ago that was getting to be a little unhealthy. The demand was so much higher than the supply that it was possible for people to enter the industry, and enter manufacturing, without even having a decent product. Whatever product they could produce they could sell at a significant profit, even if it was a poor-quality product.[35]

Wenham was speaking in April 2010, by which time Suntech had become the world's largest producer of crystalline silicon solar cells (despite being founded only nine years earlier in 2001). The rise of the Chinese producers like Suntech and Yingli over the past five years has allowed mere theory to be overtaken by a brutally real example of the way in which the lure of easy money eclipses slow and steady work on innovation. As demand for solar panels increased, there was a bottleneck in the silicon supply in 2004–5: by 2006 German production lines for solar cells were estimated to be using only 60 per cent of their capacity, a fall of over 20 per cent from the previous years.[36] As a result, the German installers turned to China for supply. Since then, Suntech, Yingli and other Chinese producers have cut the costs of solar cells by a very large margin,[37] and in 2010 Bloomberg New Energy Finance reported the inevitable results:

> China's manufacturers grabbed 43 per cent of the global photovoltaic-panel market in the last six years, pricing products as much as 20 per cent cheaper than European offerings.[38]

Consequently, Germany now imports some two-thirds of its solar PV panels from China: in essence, the PV feed-in tariff, ultimately paid for by the German consumer, is subsidising Chinese manufacturers. Understandably, this has caused unease, and the rumours in 2009 that Suntech was selling cells in the US for less than cost in order to break into the market found willing listeners in spite of official denials.[39] What cannot be doubted is the likelihood that China's market dominance will affect Germany's exports of solar cells, an impact of extraordinary importance since exports, as the *EmployRES* study described above shows at length, are widely expected to be the only certain way of increasing employment in return for renewable subsidy. Germany's solar cell industry is relatively sophisticated, and is making considerable strides in increasing the efficiency of so-called CIGS, which use a thin (and therefore cheap and less resource consuming) layer of the semi-conductor copper indium gallium diselenide. However, the technology remains at the research stage, while established but unadventurous crystalline silicon solar cells from China absorb the available subsidies.

The increasingly cheap panels have combined with the preferential tariffs to make PV installation financially appealing, and 2009 saw an extraordinary explosion of installations, representing a huge increase on the previous years:

Figure 7:8
German annual new capacity, PV installations, 2000–2010

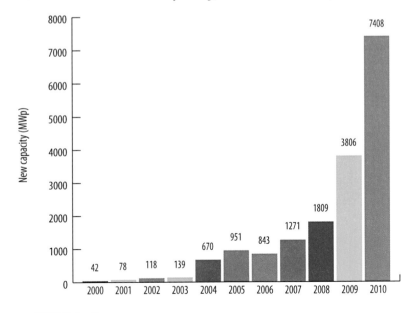

Source: BSW-Solar[40]

In the face of this unprecedented increase, on 1 January 2010 the German government increased the standard yearly degression for the PV feed-in tariff from five per cent to ten per cent, causing an outcry from the German solar cell industry. Henning Wicht from iSuppli predicted that the number of PV installations in the country would sink like a stone after April 2010. However, as Figure 7:9 shows, the upward trend in installations continued, and a second degression, this time of 16 per cent, was agreed for May 2010. Further industry protests resulted in the postponement of the degression until July, and, after a great deal of wrangling in parliament, in June the Upper House of the Bundesrat refused to pass the law as it stood, though a compromise meant that a 13 per cent cut was followed three months later by a further three per cent cut.

Figure 7:9
PV installations, January 2009–December 2010

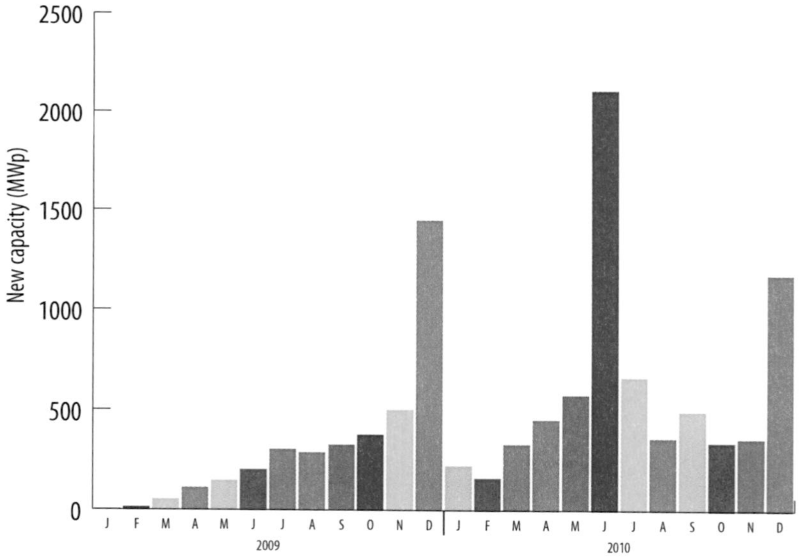

Source: Polder PV[41]

France

This influx of cheap solar panels, mainly from China, has affected other countries, for example France, which had brought in 20-year feed-in tariffs as early as March 2002. By 2006 the tariffs for built-in PV on roofs were at German levels: 30 cents/kWh for the basic PV tariff, plus a bonus of 25 cents/kWh. A year later the French Ecology Secretary Jean-Louis Borloo launched the environmental round-table, *Le Grenelle de l'environnement*. Borloo announced a target of 5,400 MW of installed PV, and acceded to industry requests to increase the PV tariff on commercial buildings to 45 cents/kWh. However, the fall in panel prices led to a rapid rise in PV installations, leading French ministers to fear a 'bubble', and there were press stories of uninhabitable house shells being built in the South of France simply to carry feed-in tariff registered panels. Like Germany, France cut its tariffs twice in 2010, and in December went further still, announcing a four-month moratorium on all planning applications for large photovoltaic installations. In the meantime, Borloo has been promoted, out of harm's way, and the *Grenelle* quietly forgotten.

Conclusion

Future developments in Germany remain uncertain, but will depend crucially on public reaction to further increases in prices resulting from subsidies to renewables and whether the net economic impact of those developments is seen as positive. With regard to the former of these matters, price tolerance, the prognosis is not good. In new work, Peter Grösche (RWI) and Carsten Schröder (University of Kiel, and DWI, Berlin) report on willingness to pay (WTP) for various fuel mixes, and conclude:

> Albeit people's WTP for a certain fuel mix in electricity generation is positively correlated to the renewable fuel share, our results imply that the current surcharge effectively exhausts the financial scope for subsidising renewable fuels.[42]

Given the heavy backloading of net economic impacts suggested by Hillebrand, it seems possible that consumer tolerance in Germany will erode more rapidly than expected, and since much governmental and public enthusiasm for their green energy support mechanisms has been premised on a successful transition to a new era of industrial prosperity for Germany, the disappointment of these expectations may have serious implications for the continuation of those policies.

Chapter Eight
Spain's Solar Eclipse

Introduction

Demand for energy in Spain has grown rapidly in recent decades for a variety of reasons, mostly if not entirely connected with economic liberalisation. Spanish GDP grew by 49.1 per cent between 1990 and 2005, while the country's primary energy use increased by 61.3 per cent.[1] Spain has modest indigenous fossil resources, and most of this demand is accounted for by imported fuels.

Figure 8:1
Primary energy use in Spain, 2008

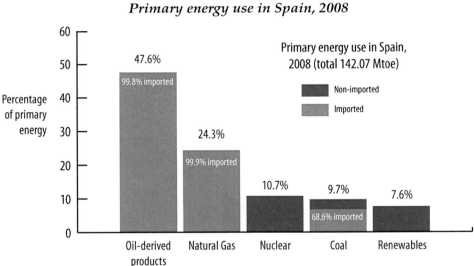

Source: 'An update of Spanish renewable energy policy and achievements in a low-carbon context'[2]

It is unsurprising that Spain should be interested in the development of renewables, quite aside from climate change policies, and the government has issued a series of legislatory mandates to increase the contribution of such energy sources, though as with all EU states these should also be seen in the context of attempts to comply with the various EU Directives on this matter. The scheme relevant to a retrospective assessment is Spain's *Plan de Energías Renovables* (Renewable Energy Plan 2005–2010), which required that 12.1 per cent of primary energy should be obtained from renewables. This would comprise 30.3 per cent of electricity generation and 5.83 per cent of transport fuels from renewable sources by 2010.[3] The policy instruments employed to drive this level of deployment in electricity, approximately 102.3 TWh of 337.4 TWh, were principally feed-in tariffs.

Spain's Renewables: Cost and Benefits

While these measures have driven a rapid growth in renewables and a significant wind power and solar PV industry in Spain, with some exports, the impact of these measures on prices to consumers has been highly significant and controversial. A study released by four researchers led by Gabriel Calzada Álvarez at the Universidad Rey Juan Carlos in 2009 indicated that the cost burden over and above the cost of electricity from conventional electricity sources between 2000 and 2008 amounted to around €8 billion.[4] Álvarez argued that this expenditure was not net positive in terms of employment creation, and even attempted to calculate to what degree this was the case, offering the suggestion that 2.2 jobs in other sectors had been destroyed for each position created in the subsidised renewable sector. This was extremely controversial, not least because it explicitly commented on a prominent speech made by President Obama citing Spain, along with Germany and Japan, as a glowing example of the green economy, and in support of his own initiatives.[5] Indeed, Álvarez explicitly warns the United States against positive expectations. This paper produced rapid responses, firstly from the Spanish trade union *Instituto Sindical de Trabajo, Ambiente y Salud* (United Institute for Employment, Health and the Environment, ISTAS),[6] which remarks on the rhetorical style of the paper, and among other points criticises Álvarez for attributing increases in Spanish electricity prices to the renewables policy, a point on which, in fact, he was supported by government analysis.

A second rebuttal was published in August 2009 by the US National Renewable Energy Laboratory (NREL).[7] This document criticises Álvarez for failing to use input-output tables when assessing the economic impact of subsidy to renewables, but instead employing two non-standard methods, best summarised in Álvarez's own words:

> In order to know how many net jobs are destroyed by a green job programme for each one that it is intended to create, we use two different methods: with the first, we compare the average amount of capital destruction (the subsidised part of the investment) necessary to create a green job against the average amount of capital that a job requires in the private sector; with the second, we compare the average annual productivity that the subsidy to each green job would have contributed to the economy had it not been consumed in such a way, with the average productivity of labor in the private sector that allows workers to remain employed.[8]

Álvarez thus notes that the average subsidy per renewables worker is €571,138 and the average capital per worker in the general economy is €259,143. He thus concludes that each green job has 'destroyed' 2.2 jobs in the Spanish economy, though he might perhaps have more cautiously described these as opportunities for wealth-creating employment forgone. NREL's fundamental criticism rests on the assertion that 'there is no justification given for the assumption that government spending (e.g. tax credits or subsidies) would force out private investment' and concludes that: 'This assumption is fundamental to the conclusion that Spain's renewable energy policy has resulted in job loss.' However, a footnote to this sentence observes that 'government spending may result in reallocation of resources', which is in fact what Álvarez set out to indicate.

Indeed, one might respond to NREL's authors that it is not at all unreasonable to assume that the impact of raising taxes is to suppress economic activity, and that if the taxes are raised by increasing electricity prices then that reduction could be felt across a wide range of economic fields. As Hillebrand and his colleagues remark in their 2006 paper on the impact of German feed-in tariff subsidies:

> In general, the electricity cost impulse will lead to contractive sectoral production and employment effects. Although the adverse effect affects primarily base production, retail, transportation and services are affected as well. The strong increase in electricity prices particularly for private households will induce reductions in real income and private consumption and, therefore, further amplify the contractive effect. In 2010, the overall economic growth will slow down by approximately 0.1 per cent; this will induce the loss of 23,000 jobs.[9]

Thus, while it is possible to quibble with Álvarez's technical approach to measuring the impact of such subsidies, the likelihood of such an impact does not require the special justification NREL describes.

Rising electricity prices

In retrospect, in spite of any arguable faults, the Rey Juan Carlos Universidad paper served to open up the question of net employment benefits and the wider economic impact of Spanish subsidy to renewables by highlighting the high costs of the Spanish support mechanisms, and on this latter point, which was contested by ISTAS, Álvarez has been vindicated. A document released by the Spanish government's Ministry of Industry, Tourism and Commerce (Ministerio de Industria, Turismo y Comércio: MITYC) in April 2010 reveals the full scale of the cost of the scheme and the impact on electricity prices.[10] This text is a key document in the discussion, and has been reported in Spain itself as being supportive of the view that the support for renewables has not been net positive for the Spanish economy.[11] Since the document is only available in Spanish there may be some value in recapitulating its arguments, and redrawing its charts with translated captions.[12]

Under the title 'The Price of Electricity Affects the Well-being of Households', the presentation observes that in the years 1998 to 2009 household electricity prices have risen from a level where they were slightly below the European average to being about five per cent above it, a 36.8 per cent increase. In the subsequent slide the presentation notes that electricity price is a major determinant of industrial competitiveness, and prices in that sector have risen significantly over the period 1998 to 2009, leaving the country with prices that are 17 per cent above the EU average. The author observes that for some industries, such as cement, industrial gases, meals, chemistry, iron and steel, energy costs are three times labour costs. While the domestic impact is serious, it is the effect on industries and their competitiveness that concerns the Ministry's analyst, and the matter is summarised in the following chart.

Figure 8:2

Historical evolution of electricity prices to industry and the general market (pool) price,
December 1996–December 2009

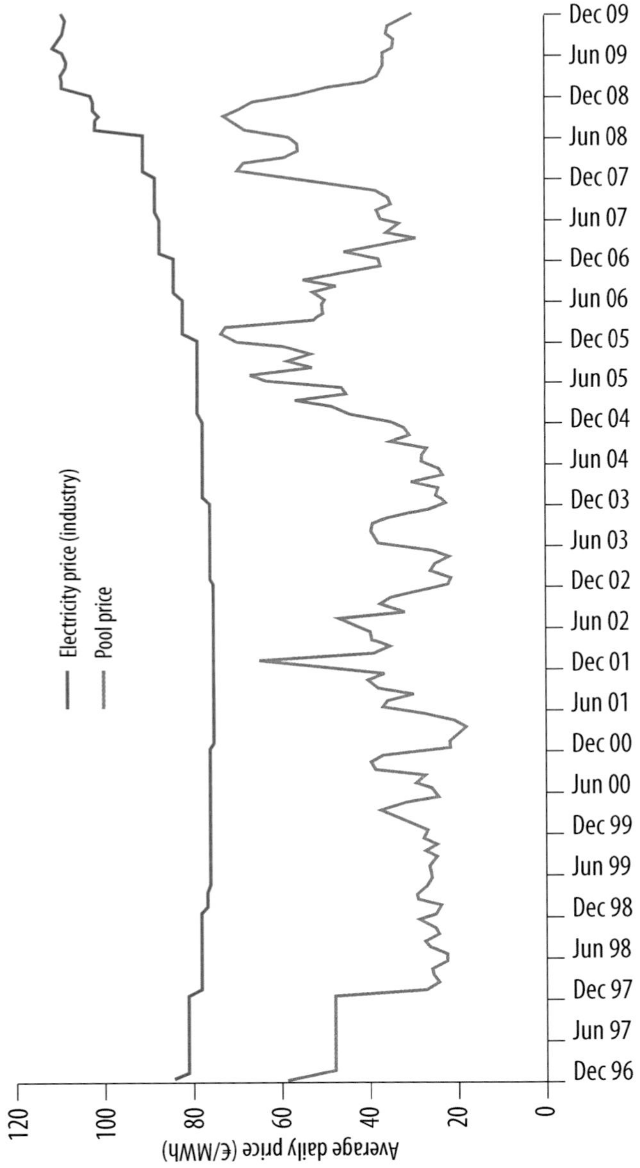

Source: 'Energías renovables: situación y objetivos'

95

The author comments in the text below this illustration that: 'The increase in industrial electricity prices cannot be explained by prices in the electricity market (pool), which have actually fallen since 2005', and in the following slide draws the explicit conclusion that '120 per cent of the change in electricity bills is explained by the increasing additional costs of renewables. This has counteracted the 25 per cent reduction in costs of conventional production.' Importantly, the author observes that these cost increases are the result of subsidy only, and do not include indirect ancillary costs that are attributable to renewables:

> The indirect costs of renewables must be added to their direct costs. For example, the necessity of additional investment in the networks to integrate renewables (which are around ten per cent of the predicted investments in planning) and capacity payments to flexible back-up installations (coal and gas) that are working fewer hours.

This is a highly significant point, on which far too little attention has currently been focused. Unfortunately it is notoriously difficult to assess these ancillary costs and decide what proportion of them are attributable to renewables policies.

Of course, such costs should be judged in relation to the benefits they return, and since a principal justification for the programmes is the diversification of Spain's fuel mix and the reduction of emissions, it is appropriate to examine these. The Ministry observes that, thanks to the increase of renewable energy in the mix, 'the level of [domestic] energy production has increased by three percentage points since 2005, to 23 per cent of total supply, and the cost of energy imports has fallen by €5.05bn (including hydro-electricity)'. Furthermore, though this will be based on debatable assumptions about fuel displacement and system effects in the electricity sector, that 'emissions have been significantly reduced, due to a much cleaner electricity generation mix (a decrease of 0.120 tonnes of CO_2 per MWh)'. These points are illustrated in the following charts.

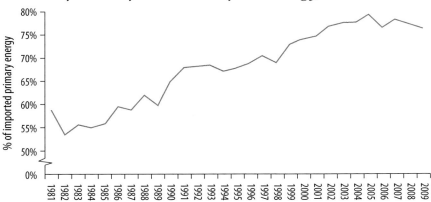

Figure 8:3
Spanish dependence on imported energy, 1981–2009

Source: 'Energías renovables situación y objetivos'

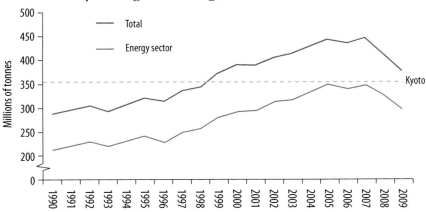

Figure 8:4
Spanish greenhouse gas emissions 1990–2009

Source: 'Energías renovables situación y objetivos'

These are significant, if modest, improvements, but both are strongly correlated with the onset of the recession, and this factor may be as important as renewables themselves. It is worth noting, for example, that UK emissions have fallen in a similar fashion, and it is widely recognised that this results as much from the economic downturn as it does from climate change policies, with the expectation that there may be a rise in emissions should the economy return to stronger growth.[13]

Figure 8:5
UK greenhouse gas emissions by source

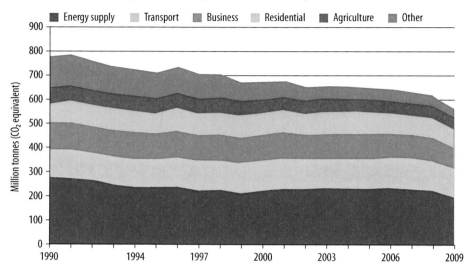

Source DECC[14]

The United Kingdom's own Department of Energy and Climate Change remarks:

> The overall decrease in emissions has primarily resulted from two factors: a significant fall in energy consumption across all sectors, and an increase in the use of nuclear power rather than coal and natural gas for electricity generation. As the UK economy contracted during 2009, this resulted in an overall reduction in demand for electricity, together with lower fossil fuel consumption by businesses and households.[15]

A similar effect is likely to be the case in Spain, with the gross reduction in emissions being largely due to economic contraction, with some related effect on import dependency. Nevertheless, as the Ministry indicates, the emissions factor of the Spanish electricity sector has improved by 0.12 tonnes per MWh, which is significant and desirable. However, not all of this improvement can be attributed to the contribution of renewables, since growth in that sector has been more than matched by significant expansion in other low-carbon generators, such as natural gas, which are also displacing coal and oil, as can be readily observed in this International Energy Agency chart:

Figure 8:6
Spanish electricity generation by fuel type, 1972–2008

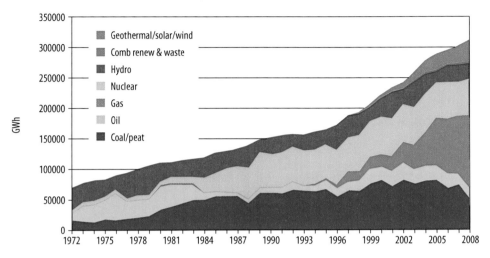

Source: International Energy Agency[16]

The economic impact of Spain's renewable subsidies

Even if we were to attribute all of the emissions savings and reduction in import dependency to the renewables policy, it is not clear that this represents value for money, and the savings should be put into context with the costs currently incurred, and future costs that are implied by current policies. The Ministry's text observes that between 2004 and 2010 subsidies increased by a factor of five, and that: 'In 2009 alone, the amount of subsidies doubled from the previous year, reaching €5.045 bn, the equivalent of all the public investment in research and development and technological innovation in Spain'. The rapidity of the cost growth is in itself an interesting feature of the topic, and is clearly visible in the Ministry's chart:

Figure 8:7
Growth and projected growth in cost of subsidy to Spanish renewable electricity generators, 2004-2012

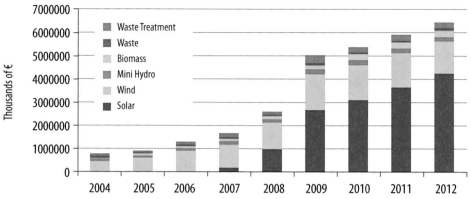

Source: 'Energías renovables situación y objetivos'

Even if no further renewables are installed, those currently operating 'will receive more than €126bn over the next 25 years', a figure that, as the government itself notes, can only rise still further in view of Spain's continuing commitment to the EU Renewables Directive. It is interesting to observe that while solar accounts for 11 per cent of renewable energy in 2009, it is responsible for 53 per cent of the costs. Wind power, by comparison, provides 64 per cent of the energy from renewable sources, but is responsible for 31 per cent of the costs. This leads the Ministry to observe that wind power in Spain has also generated €1.3bn in exports, and saved €3.6bn in avoided fossil fuel imports, though it is unclear how this figure has been calculated. On the other hand, 'around 62 per cent of PV cells and modules were imported at a cost of €5.182bn, equivalent to 28.6 per cent of net imports of crude oil and derivatives in 2007'.

While some components of the Spanish renewable policy have been less expensive than others, the net economic impact of these policies has been to increase costs and leave the Spanish government with a significant debt, the so-called Tariff Deficit. This has occurred since, in certain regulated parts of the market, generators are not permitted to pass on the full cost of generation and supply to customers, the difference being accumulated on their balance sheets as government debt. This amounted to some €16.5bn at the end of 2010, over half of which was owed to one company, Endesa.[1] The evolution of this debt is shown in MITYC's 2010 document:

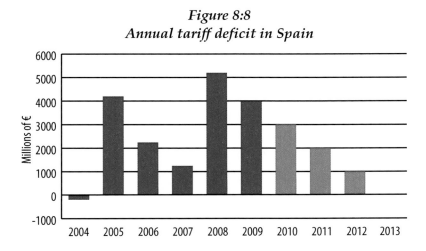

Figure 8:8
Annual tariff deficit in Spain

Source: 'Energías renovables situación y objetivos'

The government is committed to eradicating the deficit by 2013, but the government's text admits that, in spite of its efforts, this is proving difficult. The alternatives before the government include allowing the creditor utilities to collect the deficit through increased charges to consumers, which will be unpopular, and the conversion of the debt to government-backed fixed-income securities, but as one market analysis remarks: 'When the deficit is securitised, bonds that are issued will compete with Spain's sovereign debt.'[18]

Conclusion

However this problem is resolved, the Spanish government is faced with a necessary reduction in tariff rates for renewable generators, particularly solar photovoltaic. The feared economic impact of such revisions can be judged from the headlines reporting these developments in late 2010: 'Spain's Solar Deals on Brink of Bankruptcy as Subsidy Policies Founder'.[19] These changes are highly controversial, entailing as they do not only reduced income for existing projects, but significantly contracted plans for expansion. One industry journal reported the topic under the headline 'Spain's Solar Power Sector Falls into the Abyss'.[20] The government appeared determined, however, and just before Christmas passed a royal decree to cut subsidies by 30 per cent, a decision accepted by the congress, in spite of lobbying, at the end of January 2011. However, the industry has responded with threats of legal action, and the situation remains unresolved.[21]

These extraordinary developments make it plain that the solar explosion, so amply described in the Ministry's publication, was, as Álvarez argued in his controversial text, a classic subsidy-driven bubble. In an interview with an environmental journalist for Ecoseed in early 2011, a spokesman for the industry body ASIF (Asociación de la Industria Fotovoltaica) remarked 'The government cheated the solar investors by changing the law after it has lured them to invest their money in PV power plants... If you know that the government would change the law, you will never have invested in that technology and never have put your money in that market'.[22] This implicitly concedes that the sector was from the outset likely to be a long-term client of the state, unable to survive without support, and should serve as a warning to other governments hoping to create independent renewables industries through subsidy.

In retrospect, while we cannot conclusively validate the precise calculations of employment effects in Álvarez's assessment, it is clear that the government's substantial reallocation of resources within the Spanish economy produced consequences that were soon found to be intolerable both in the short term and in view of implied long-term costs. It is reasonable to infer that government's reluctance to continue these subsidies is grounded in a recognition that they are a brake on overall economic recovery at a time when a return to growth is urgently required.

Conclusion

Green Prospects in the UK

Whatever the longer-term potential for a viable and prosperous global economy with a low-emissions profile, this study shows that the prospects for a self-sustaining low-carbon economy as the result of current UK national and EU-wide policies are poor. Since these policies imply high levels of state coercion, with the risk of stagnating growth and low levels of invention and innovation, they would also appear to be a weak preparation for a period of fossil fuel resource erosion.

Historical examples of peacetime state management on the large scale, such as the United Kingdom's East African Groundnut Scheme, are extremely discouraging, and even the New Deal, a major inspiration for proponents of the low-carbon industrial policy, is now being reassessed in ways that render it less satisfactory as a precedent for action. Such matters give no encouragement to governments hoping that a target-driven transition to renewables will be an engine of recovery. Failure is not unlikely, and even success may be counterproductive.

Furthermore, empirical experience in Spain and Germany shows that the costs of supporting renewable energy generation is high, compared to low-carbon alternatives, and almost certainly has, over time, net economic effects that are negative both in terms of GDP and employment.

Analysis for the European Union suggests that the net effects of such policies will only be positive, and then but marginally so, if the EU retains a high share of the world export market in renewable energy technologies. This seems improbable in the light of the rapid growth of renewable energy exports from China in the last decade, and the likelihood of comparative advantages in other developing economies, not to mention advanced OECD states outside the EU. Furthermore, even if the EU can retain control of the export markets, it appears that such markets are dependent on the continuation of consumer-derived subsidies in the purchasing countries, and experience within the European Union suggests that this is not likely.

Perhaps most distressing of all, the EU's analysis also shows that even in scenarios which are net positive for the EU as a whole some member states would probably suffer reduced growth and employment loss, with the UK foremost amongst these 'loser' countries. This may seem surprising, since international engineering firms are showing interest in constructing offshore wind turbines in the UK: for example, the Spanish company Gamesa has announced its intention to build an offshore wind technology centre in Scotland, creating 130 jobs,[1] while Siemens is planning an ambitious proposal at Hull, which it is hoped will provide up to a thousand permanent jobs. This has prompted local politicians to declare a 'new era of prosperity for East Hull',[2] and the *Guardian* to describe it as 'a shot in the arm for the government's attempts to create new jobs from the "green economy"'.[3] However, it must be recognised that these positions are dependent on demand created by public expenditure through the Renewables Obligation, and while the employment effect both locally and nationally may be positive, in the investment phase it is probable that this will change as the cost of electricity rises as a consequence of the subsidies. Hull's gain will be somewhere else's loss, and ultimately a loss overall for the UK economy. Furthermore, judging from the sensitivity of net benefit to trade in renewable technology, as revealed by the EU's *EmployRES* study, both instances are probably best seen as bridgeheads to secure exports from Germany and Spain into the UK market.

Indeed, the UK overall is still in the investment phase with regard to renewables, but the on-costs are beginning to be evident. The principle instrument for encouraging renewable energy development since 2002 has been the Renewables Obligation, a complex system of indirect subsidy that places an obligation on suppliers to purchase from renewable generators a set percentage of the electricity they sell to customers. Compliance is demonstrated by the presentation of Renewable Obligation Certificates, which are often purchased from the generator with the electricity, but can be traded separately. Failure to comply results in a fine, which is paid to the regulator, Ofgem, with these fines then being distributed proportionally to those suppliers who have demonstrated compliance. The value of the Renewables Obligation Certificate is thus twofold: the value of the fine avoided, and the expected share of the fines paid by others. The approximate total value of the certificate during the life of the scheme has varied little, and £50 is often used in industry calculations.[4] This may be compared with a wholesale electricity price that has varied over the life of the scheme from approximately £25 to over £40 at present. Thus a renewable generator may be expected to receive £75–£100/MWh generated, a premium of approximately 50 per cent over the wholesale price. Suppliers recover this additional cost from consumers; in other words, from the supplier's perspective, the Renewables Obligation is simply a pass-through cost to consumers. The Treasury understandably classifies the Renewables Obligation as a tax, and the transfer of monies to renewable generators as public expenditure.[5]

These generous returns have encouraged considerable and rapid deployment up to 2010, as can be seen in the following chart generated from the Renewable Energy Foundation database, which is based on corrected and reprocessed Ofgem Renewables Obligation data.[6] The chart also graphs projected growth in keeping with current UK government projections of the installed capacities of wind power needed to meet targets (growth in other sectors is expected to be small).[7]

The empirical portion of the chart, up to 2010, shows fairly rapid growth from a low base. A very rapid increase is required to meet the 2020 targets and entails the generation of approximately 75 TWhs, which on government predictions would be about 25 per cent of the quantity of electricity supplied to consumers in 2020. Use of historical ROC prices enables us to calculate the subsidy cost to consumers of the dedicated renewables capacity (i.e. excluding co-firing, which use existing coal plants) since the beginning of the scheme, and to assess the likely future cost of plant constructed for the scheme.

The empirical portion of the chart shows that the cost to date, from 2002 to 2010, amounts to approximately £5.6 billion, with the oncost to 2020 adding a further £39 billion, a sharp increase in consumer burden. If we assume that after 2020 no further efforts are made to expand capacity, but that subsidies are maintained for capacity already installed under the RO, a further £60 billion would be added to bills, with the result that the total cost of the scheme from 2002 to 2030 would amount to £100 billion. This is clearly a high cost, and will have a significant effect on domestic and industrial electricity prices. The Department of Energy and Climate Change's own

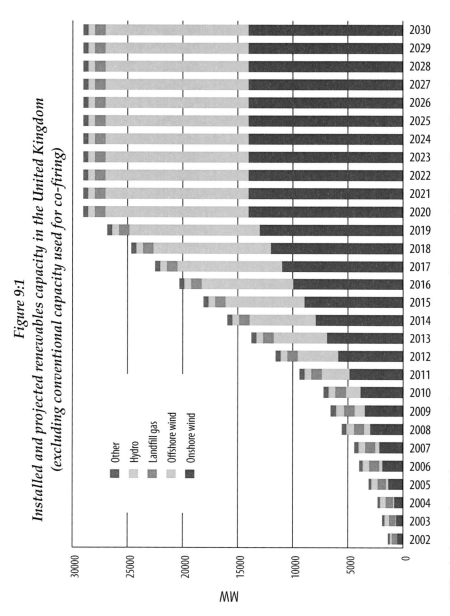

Figure 9:1
Installed and projected renewables capacity in the United Kingdom
(excluding conventional capacity used for co-firing)

Source: REF calculations from Ofgem empirical data and Department of Energy and Climate Change Projections

Figure 9:2

Cost and projected cost of the Renewables Obligation to UK consumers

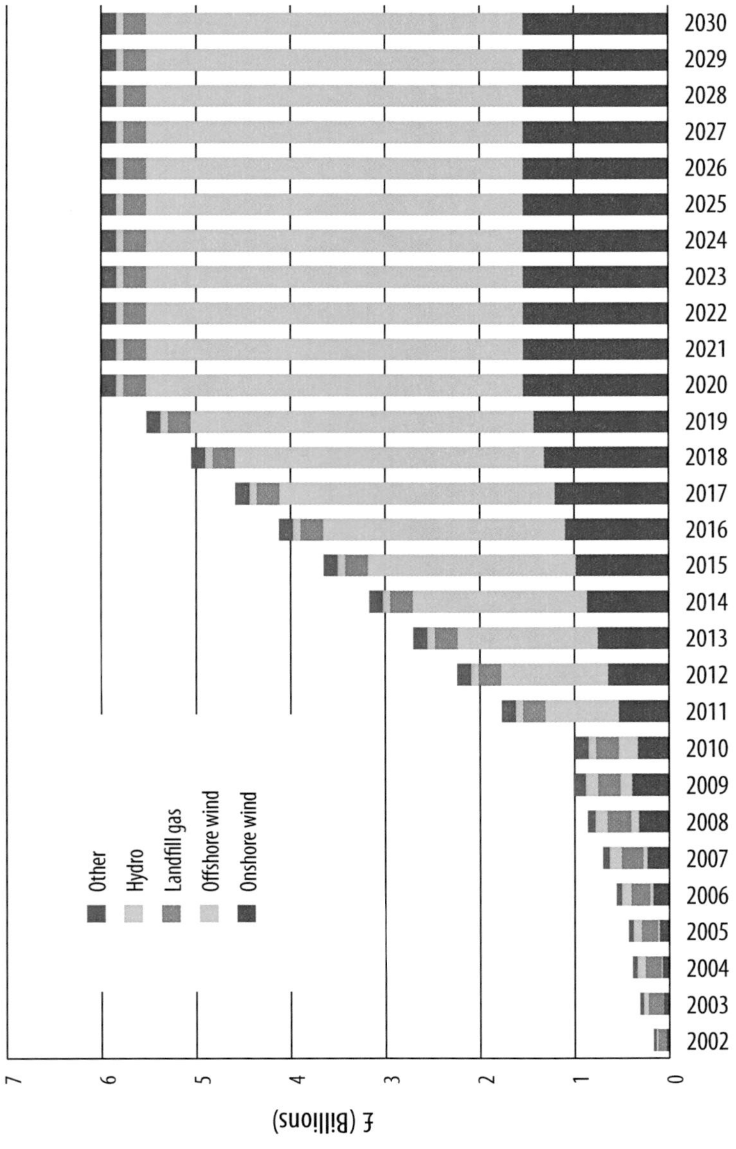

Source: REF calculations from empirical Ofgem data. Projected costs assume a ROC price of £50

calculations assume that it will add £94 a year (2009 prices) to the average domestic electricity bill in 2020, and £210,000 annually to the average medium-sized non-domestic energy user's bill.[8] The Department has not issued estimates of the impact on energy-intensive and large business users.

Expenditure on this scale has, of course, had a gross economic effect in terms of employment, and the UK wind industry association has recently published a study, *Working for a Green Britain: Employment and Skills in the UK Wind & Marine Industries* (2011), that helpfully estimates the levels of full time equivalent (FTE) employment in these sectors.[9] The total number of FTE positions is put at 10,800 in 2009/10, with 56 per cent of these being in onshore wind of greater than 100 kW, 29 per cent in offshore wind, seven per cent in microwind, and eight per cent in marine and tidal.[10] The study goes on to note that the 9,200 FTE positions in wind in 2009/10 compares with a figure of 4,800 in 2007/08. This leads RenewableUK (REUK, formerly BWEA) to claim 91 per cent growth in full-time employment in wind between these two periods and to observe that: 'The growth in employment stands in contrast to the overall UK employment level, which has shrunk by 3.4 per cent'.[11] Maria McCaffery, the trade lobby's CEO, comments:

> Two conclusions from the results of this remarkable study are immediately obvious: this sector has withstood the negative GDP growth of the UK recession and bucked the overall employment trend in a spectacular way by a near doubling of the workforce.[12]

However, it is not surprising that a heavily subsidised sector should be immune to perturbations in the wider economy. Indeed, due to the fact that the Renewables Obligation subsidies impose burdens on consumers, it contributes to the pressures causing economic contraction and employment loss, though it is hard to estimate how much of the 3.4 per cent fall in employment is due to the over £1 billion annual cost of the Obligation.

We can, however, estimate the subsidy cost of establishing the 9,200 FTE positions in wind power by calculating the total subsidies paid to wind during the relevant periods, using Ofgem data, and then dividing by the number of positions. This calculation reveals that the total subsidy per FTE worker in the wind industry in the period April 2002 to March 2010 amounted to approximately £200,000. Subsidy per worker in the year April 2009 to March 2010 amounted to approximately £57,000 per FTE worker, which is greatly in excess of the median earnings in either the public (£29,000) or the private sectors (£25,000).[13]

It should be noted that this estimate is of *subsidy* per job, and does not include other increases in the cost of electricity that result incidentally from the policies, such as grid expansion, system balancing and the cost of running residual plants at a low-load factor. Such figures are notoriously difficult to estimate with confidence, but are thought to be significant.

Even though it is probably an underestimate of overall economic impact, the subsidy cost of wind industry jobs does, however, make it clear that, even if successfully implemented, the current low-carbon agenda would require that the energy sector takes up a much larger share of the overall economy than is currently the case. Since this sector is almost entirely dependent on state mandates for its income and operation, it would in effect enjoy a state-approved monopoly, with the rents being shared with the labour force at the expense of the rest of the economy, where there would be net job losses.

This does not appear to be a recipe for a prosperous or a stable society. However, these longer-term anxieties are likely to remain purely theoretical concerns, since the low-carbon jobs agenda as currently envisaged will in the medium term almost certainly fall foul of technical failure or extreme costs and consequent reduced growth, leading to distressed policy correction. Major improvements in the cost-efficiency of renewable energy generation might change this prospect, though this would entail, amongst other cost reductions, fewer employees per kWh produced. However, current policies mitigate against dramatic technological improvement of this kind.

In the absence of such a shift, only rising fossil fuel prices seem likely to obviate the need for mandates and public support mechanisms to ensure a counter-market shift to renewable energy. However, even if fossil sources do become more expensive, resulting in a market-driven transition, the fact remains that unless renewables achieve levels of productivity comparable with the current fossil fuel industry, a painful rebalancing of the economy will occur, with the bulk of our societal resources being cycled within the energy sector, leading to corresponding reductions of prosperity elsewhere in the economy. An age of subsistence energy generation will have dawned.

Notes

Chapter One: The 'Triple Win'

1 http://www.ilo.org/integration/events/events/lang—en/WCMS_119192/index.htm

2 David Cameron, speech to the Confederation of British Industry (CBI), 25 January 2010: http://www.number10.gov.uk/news/speeches-and-transcripts/2010/10/creating-a-new-economic-dynamism-56115.

3 Stavros Dimas, 'What Jobs in a low-carbon Economy', Speech 21 February 2007. http://europa.eu/rapid/pressReleasesAction.do?reference=SPEECH/07/92

4 'EU Businesses lead move to a low-carbon economy – Conclusions of the High Level Group on Competitiveness, Energy and Environment',Brussels,30.11.07. http://europa.eu/rapid/pressReleasesAction.do?reference=IP/07/1823&format=HTML&aged=0&language=EN&guiLanguage=en

5 Steven Greenhouse, 'Millions of Jobs of a Different Collar', *New York Times*, 26.03.08. http://www.nytimes.com/2008/03/26/business/businessspecial2/26collar.html

6 Greg Barker, MP, Minister of State for Climate Change, 'Your Green Economy Needs You', *Business Green*, 8 February 2011. Available from the DECC website: http://www.decc.gov.uk/en/content/cms/news/GB_BizGreen/GB_BizGreen.aspx.

7 http://www.campaigncc.org/

8 See www.climate-change-jobs.org.

9 Campaign against Climate Change's pamphlet *One million climate jobs: solutions to the economic and environmental crises*, London: CaCC, 2010, p. 8. A report by the Campaign against Climate Change trade union group in conjunction with the Communication Workers Union (CWU), the Public and Commerical Services Union (PCS), the Transport Salaried Staffs Association (TSSA), and the University and College Union (UCU). Available from www.climate-change-jobs.org.

10 *One Million Climate Jobs*: Technical Note: Jobs Gained and Lost, 2010, p. 4. Only available online. See www.climate-change-jobs.org/node/14.

11 *One million climate jobs: solutions to the economic and environmental crises*, p. 8. A report by the Campaign against Climate Change trade union group in conjunction with the Communication Workers Union (CWU), the Public and Commerical Services Union (PCS), the Transport Salaried Staffs Association (TSSA), and the University and College Union (UCU). Available from www.climate-change-jobs.org.

12   *One Million Climate Jobs*: Technical Note: Jobs Gained and Lost, p. 6.

13   http://www.ic.nhs.uk/statistics-and-data-collections/workforce/nhs-staff-numbers/provisional-monthly-nhs-hospital-and-community-health-service-hchs-workforce-statistics-in-england

14   *One Million Climate Jobs*, p. 9.

15   National Audit Office, *National Health Service Landscape Review*, 2011, p. 7.

16   Association of the British Pharmaceutical Industry, *Understanding the 2005 PPRS*, 2005, p. 4.

17   Wordwatch Institute for UNEP, *Green Jobs*, 2008, p.3.

18   *Green Jobs*, p. 3.

19   *Green Jobs*, p. 4.

20   *Green Jobs*, p. 4.

21   *Green Jobs*, p. 4.

22   *Green Jobs*, p. 6.

23   *Green Jobs*, p. 6.

24   *Green Jobs*, p. 8.

25   *Green Jobs*, p. 10.

26   *Green Jobs*, p. 13.

27   *Green Jobs*, p. 24.

28   *Green Jobs, p. 24.*

Chapter Two: Embracing Wealth Destruction

1   Simms, A., *The New Home Front*, House of Commons: Caroline Lucas MP 2011. Available from http://www.carolinelucas.com/cl/media/the-new-home-front-uk-needs-a-war-footing-on-energy-and-climate-crisis.html Green New Deal Group, *A Green New Deal: Joined-up policies to solve the triple crunch of the credit crisis, climate change and high oil prices*, London: New Economics Foundation, 2008. Available from: http://www.neweconomics.org/projects/green-new-deal.

2   *A Green New Deal*, p. 16.

3   *The New Home Front*, p. 30.

4   Barker, G., 'Your Green Economy Needs You', *Business Green*, 8 February 2011. Available from the DECC website: http://www.decc.gov.uk/en/content/cms/news/GB_BizGreen/GB_BizGreen.aspx.

5   *A Green New Deal*, p. 38. *The New Home Front*, p. 20.

6   *The New Home Front*, p. 9.

7   *A Green New Deal*, p. 33.

8   *The New Home Front*, p. 13.

9   *The New Home Front*, p. 4.

10   *The New Home Front*, p. 13.

11   *The New Home Front*, p. 13.

12   *The New Home Front*, p. 13.

13   *A Green New Deal*, p. 32.

14   Marks, N., *The Happiness Manifesto*, Kindle edition: TED Books, 2011.

15   For a review see Oswald, A., 'Emotional Prosperity and the Stiglitz Commission', forthcoming in the *British Journal of Industrial Relations*, December 2010. Available from http://www.andrewoswald.com/.

16   *The New Home Front*, p. 9.

17   Matheson, J., *The UK Population: How does it compare?*, London: Office of National Statistics, 2010, pp. 19–21: http://www.statistics.gov.uk/articles/population_trends/03-poptrends142ns.pdf

18   *The New Home Front*, p. 14.

19   *The New Home Front*, p. 14.

20   *The New Home Front*, p. 29.

21   *A Green New Deal*, p. 29.

Chapter Three: Renewable Jobs in the EU

1   See Department of Business, Enterprise and Regulatory Reform (BERR), 'Draft Options Paper on Renewables Target', August 2007. http://image.guardian.co.uk/sys-files/Guardian/documents/2007/08/13/RenewablesTargetDocument.pdf. The document cites EU-wide total annual costs to the EU in 2020 as around €24bn assuming international trading in emissions certificates (see Table 4), the BERR officials note that they regarded this as an underestimate. The costs to the UK of a 15% target for Final Energy Consumption (the level eventually adopted for the UK) are estimated by BERR (in Table 3) at between £8.7bn and £10.8bn in 2020, which is between 40 and 50% of the EU wide total cost. (Note that Table 3 transposes the figures for the 14% and 15% target cost estimates.)

2   See: http://eur-lex.europa.eu/LexUriServ/LexUriServ.do?uri=
COM:2011:0031:FIN:EN:PDF

3   Fraunhofer ISI, Ecofys, Energy Economics Group (EEG) Austria, Rütter + Parter
Socioeconomic Research + Consulting (Switzerland), Société Européene
d'Économie (SEURECO) France, Inga Konstantinaviciute (LEI) Lithuania,
*EmployRES: The Impact of Renewable Energy Policy on Economic Growth and
Employment in the European Union* (27 April 2009). Study for DG TREN, Contract
TREN/D1/474//2006. Summary and main text downloadable from:
http://ec.europa.eu/energy/renewables/studies/renewables_en.htm

4   Summary, *EmployRES*, p. 4.

5   Summary, *EmployRES*, p. 4.

6   Summary, *EmployRES*, p. 4.

7   See Constable, J., & Moroney, L., 'Low Wind Output in 2010':
http://www.ref.org.uk/publications/217-low-wind-power-output-2010.

8   Summary, *EmployRES*, p. 4.

9   For outline see Summary, *EmployRES*, p. 4; but for details of percentage share
see the main study, *EmployRES*, p. 124.

10   *EmployRES*, p. 125.

11   For a description of the matrix see *EmployRES*, p. 126.

12   Summary, *EmployRES*, p. 7. See also Figure 18, which gives in addition the net
effects for 2010 and 2030. For the relevant section of the main text see pp. 127ff,
where it is explained that the gross employment figures are derived from the
NEMESIS model.

13   Summary, *EmployRES*, p. 7.

14   *EmployRES*, p. 140.

15   Summary, *EmployRES*, p. 4.

16   Summary, *EmployRES*, p. 26.

17   Summary, *EmployRES*, p. 25.

18   *EmployRES*, p. 135. The economic model employed is NEMESIS.

19   *EmployRES*, pp. 156-185.

20   *EmployRES*, p. 159.

21   *EmployRES*, p. 162.

22   *EmployRES*, p. 182.

23   Summary, *EmployRES*, p. 4.

24 *EmployRES*, Summary, p. 6. See also the main study, p. 151.

25 *EmployRES*, p. 185.

26 *EmployRES*, p. 164.

27 *EmployRES*, p. 184.

28 Summary, *EmployRES*, p. 27.

29 Summary, *EmployRES*, p. 24.

30 For a definition of load factor, see Chapter Seven, p. xx.

31 Eirgrid (Jonathan O'Sullivan), 'Facilitating the Transition to a More Competitive, Sustainable, and low-carbon Electricity Future'. Presentation to the Irish Renewable Energy Summit, 20 January 2011, Dundalk. Discussed in Renewable Energy Foundation, 'Low Wind Power Output 2010': http://www.ref.org.uk/publications/217-low-wind-power-output-2010.

32 In addition to REF, 'Low Wind Power Output in 2010', see Michael Laughton, 'Power Supply Security with Intermittent Sources: Conventional Plant Capacity Requirements', *Power in Europe*, 460 (10 October 2005).

Chapter Five: the Green Deal and the New Deal

1 This chapter, and the next, are digressions through the hinterland of assumptions surrounding much thought about the low-carbon economy. Readers more narrowly focused on renewables policy can move directly to Part 3, Chapter 7.

2 http://www.decc.gov.uk/en/content/cms/news/pn10_104/pn10_104.aspx

3 http://www.decc.gov.uk/en/content/cms/news/pn10_104/pn10_104.aspx

4 Barker, G., Speech to the Institute of Economic Affairs, 27 October 2010.

5 http://www.decc.gov.uk/en/content/cms/what_we_do/consumers/green_deal/green_deal.aspx

6 Barker, G., In response to public questioning from the author, Institute of Economic Affairs, 27 October 2010.

7 Redrawn from Iida, T., 'Japan: New Policies to Spark Growth?', Institute for Sustainable Energy Policy, Presentation to estec2009, Munich, Germany, 25-26 May 2009, p. 2. See also Iida, T., 'Solar Thermal Policy and Market in Japan', Tokyo: Institute for Sustainable Energy Policy, 20 June 2007), p. 2.

8 McKibbin, R., 'Britain in the 1950s: Consensus or Conflict?', address to the British Academy, 19 February 2008.

9   See Department of Work and Pensions, *Transforming Britain's labour market: Ten years of the New Deal* , London: DWP, 2008. http://www.dwp.gov.uk/docs/pmnewdeal2-01-08.pdf

10   See Hinsliff, G., 'Brown unveils plan to create 100,000 jobs', *Guardian*, 4 January 2009. http://www.guardian.co.uk/politics/2009/jan/04/gordon-brown-employment-new-deal

11   A recent article by Mr Barker refers to those engaged in rebuilding the UK's energy infrastructure as a ' low-carbon army', a clear verbal echo of the Green New Deal group's 'carbon army'. See Barker, G., 'Your Green Economy Needs You', *Business Green*, 8 February 2011. Available from the DECC website: http://www.decc.gov.uk/en/content/cms/news/GB_BizGreen/GB_BizGreen .aspx; and Green New Deal Group, *A Green New Deal*, London: New Economics Foundation, 2008, p. 3.

12   *A Green New Deal*, pp. 1, 2.

13   *A Green New Deal*, p. 2.

14   *A Green New Deal*, p. 3.

15   *A Green New Deal*, p. 3.

16   *A Green New Deal*, p. 3.

17   *A Green New Deal*, pp. 3, 24.

18   *A Green New Deal*, p. 4.

19   Leuchtenberg, W. E., *Franklin D. Roosevelt and the New Deal, 1932–1940*, New York: Harper, 1963.

20   Friedman, M. & Jacobson Schwartz, A., *A Monetary History of the United States*, Princeton: Princeton University Press, 1963.

21   Rauchway, E., *The Great Depression & The New Deal: a very short introduction*, Oxford: Oxford University Press, 2008.

22   Senator Robert Wagner, speaking in the Spring of 1931, quoted in *The Great Depression & The New Deal*, p. 23.

23   *The Great Depression & The New Deal*, p. 5.

24   Krugman, P., 'A Tale of Two Moralities', *New York Times*, 13 January 2011. http://www.nytimes.com/2011/01/14/opinion/14krugman.html

[25] Cole, H. L. & Ohanian, L. E., 'The Great Depression in the United States from a Neoclassical Perspective', *Federal Reserve Bank of Minneapolis Quarterly Review* 23/1 (Winter 1999), pp. 2–24. Cole, H. L. & Ohanian, L. E., 'New Deal Policies and the Persistence of the Great Depression: A General Equilibrium Analysis', *Journal of Political Economy*, University of Chicago Press, vol. 112(4) (August 2004), pp. 779-816.

[26] Ohanian, L. E., 'What – or Who – Started the Great Depression?', *Journal of Economic Theory* 144/6 (Nov. 2009), pp. 2310–2335. Preprint freely downloadable from http://www.econ.ucla.edu/people/defaultpapers.cfm?NAME=Ohanian.

[27] A convenient introduction to this body of work can be found in Professor Ohanian's online lecture for the Jacob Marschak Interdisciplinary Colloquium on Mathematics in the Behavioural Sciences, UCLA, 8 January 2010: http://www.youtube.com/watch?v=d_YMR1Gk2JU

[28] 'What – or Who – Started the Great Depression?', Figure 1.

[29] 'The Great Depression in the United States from a Neoclassical Perspective', Figure 2.

Chapter Six: 'Operation Groundnuts'

[1] Wood, A., *The Groundnut Affair*, London: Bodley Head, 1950. Written while the Groundnut Scheme was still in operation by the former Head of the Information Division of the Overseas Food Corporation (which ran the Scheme from 1948), this is still the only full-length account published on the subject.

[2] Myddelton, D. R., *They Meant Well*, London: Institute of Economic Affairs, 2007.

[3] Hogendorn, J. S. and Scott, K. M., 'Very Large-Scale Agricultural Projects: The Lessons of the East African Groundnut Scheme', in Rotberg, R. I., ed, *Imperialism, Colonialism and Hunger: East and Central Africa*, Lexington: Lexington Books, 1983, Frankel, S. H., 'The Kongwa experiment: lessons of the East African groundnut scheme', in *The Economic Impact of Under-developed Societies*, Cambridge MA: Harvard University Press, 1955, and Gourou, P., '"Le Plan des Arachides"', *Les Cahiers d'Outre-Mer*, VII (1955).

[4] Matteo, R., 'What was left of the groundnut scheme?' in *Journal of Agrarian Change*, 6 no 2 (2006), based on his unpublished 2004 PhD thesis, *The Groundnut Scheme Revisited*.

[5] Morgan, D. J., *The Official History of Colonial Development* (Volume 2: *Developing British Colonial Resources, 1945–1951*), Basingstoke: Macmillan, 1980, Havinden, M. and Meredith, D., *Colonialism and Development*, London: Routledge, 1993 and John Iliffe, *A Modern History of Tanganyika*, Cambridge: Cambridge, 1979.

6   Figure from 'The Lessons of the East African Groundnut Scheme', 174. This seems to have been the greatest extent in the Scheme's history, and should not be confused with the total area cleared (as much as 440 km², or 120,000 acres).

7   See *They Meant Well*, p. 83. Myddleton takes the 'end' to be the handover from the Overseas Food Corporation (OFC) to the Tanganyika Agricultural Corporation (TAC) in 1954 and calculates the net cost of £46 million by assuming that of the £49 million spent by the Government, there were capital assets worth around £3 million. Other commentators put the total around £35 million, this being the approximate losses of the OFC to 31 March 1951, which the Government wrote off. This has been estimated as equal to the Tanganyika Government's total expenditure between 1946 and 1950: *A Modern History of Tanganyika*, p. 441.

8   Thompson, G., *NHS Expenditure in England* SN/SG/724, London: House of Commons Library, 2009.

9   *They Meant Well*, p.78.

10 Robertson, F., 'We investigate the Groundnuts Scandal', *Picture Post* Vol 45 No 8 (19 November 1949), p. 21.

11 'The Lessons of the East African Groundnut Scheme', p. 187.

12 'We investigate the Groundnuts Scandal', p. 18.

13 'The Lessons of the East African Groundnut Scheme', p. 168.

14 The Overseas Food Corporation in fact had two schemes to oversee: the East African groundnuts were by far the larger; the other was a relatively successful scheme, originally covering 80 km², for growing sorghum for pig-feed in Australia.

15 Cabinet Papers: CP (49) 10, 19 October 1949.

16 'We investigate the Groundnuts Scandal', p. 13.

17 'The Future of the Overseas Food Corporation', Cmd 8125, January 1951.

18 'The Future of The Overseas Food Corporation', Cmd 9158, May 1954.

19 'We investigate the Groundnuts Scandal', p. 18.

20 *The Groundnut Affair*, p. 46.

21 *Developing British Colonial Resources*, p. 210.

22 *The Groundnut Affair*, p. 36.

[23] 'A Plan for the Mechanized Production of Groundnuts in East and Central Africa', Cmd 7030, February 1947, pp. 26, 43. In the acknowledgements, where he is thanked (p. 48), he is referred to as 'Tom Bains'.

[24] From John Iliffe, p. 422; quoted in *They Meant Well*, p. 72.

[25] 'A Plan for the Mechanized Production of Groundnuts in East and Central Africa', p. 55. According to D. J. Morgan, there was a 'cleavage of opinion' about the scheme's viability within the Colonial Office: *Developing British Colonial Resources*, p. 240.

[26] See http://www.cartoons.ac.uk.

[27] *Developing British Colonial Resources*, p. 238.

[28] *Developing British Colonial Resources*, p. 306.

[29] 'A Plan for the Mechanized Production of Groundnuts in East and Central Africa', pp. 3–4.

[30] *Developing British Colonial Resources*, p. 238.

[31] *A Modern History of Tanganyika*, p. 441.

[32] 'The Lessons of the East African Groundnut Scheme', pp. 167–8.

[33] Hansard: HC Deb 18 July 1950 vol 477 cc2042.

[34] *Developing British Colonial Resources*, p. 251.

[35] Hansard: HC Deb 06 November 1947 vol 443 cc2033.

[36] According to MP Godfrey Nicholson. Hansard: HC Deb 18 July 1950 vol 477 cc2107.

[37] *The Groundnut Affair*, p. 93.

[38] A familiar expression quoted by Alan Lennox-Boyd. Hansard: HC Deb 11 March 1948 vol 448 cc1489.

[39] Hansard: HC Deb 11 March 1948 vol 448 cc1505.

[40] *Colonialism and Development*, p. 280.

[41] D. J. Morgan points out in *Developing British Colonial Resources* that the UK Government was under pressure from the US to make more use of the country's colonies.

[42] *Developing British Colonial Resources*, p. 282.

[43] In 1948, 47 per cent of world sisal production was in Tanganyika; before that year price controls meant that British Government was able to buy sisal at a low price and sell it on to the US for as much as twice the price. See *A Modern History of Tanganyika*, 344.

[44] Hansard: HC Deb 29 July 1947 vol 441 cc356–7.

[45] Hansard: HC Deb 14 March 1949 vol 462 cc1747.

[46] Hansard: HC Deb 07 March 1947 vol 434 cc873. A threefold rise in the international price of groundnuts over three years allowed Strachey to claim in 1949 that despite consistently low yields, things were only going to get better.

[47] A T P Seabrook, 'The Groundnut Scheme in Retrospect', *Tanganyika Notes and 78* 'A Plan for the Mechanized Production of Groundnuts in East and Central Africa', pp. 3–4.

[48] *The Groundnut Affair*, p. 158.

[49] 'The Groundnut Scheme in Retrospect', pp. 88–9.

[51] 'We investigate the Groundnuts Scandal', p. 15.

[52] Hansard: HC Deb 29 July 1947 vol 441 cc344.

Chapter Seven: Germany's Cloudy Future

[1] Frondel, M., Ritter, N., Schmidt, C. M., Vance, C, 'Economic Impacts from the Promotion of Renewable Energies: The German Experience. *Energy Policy* 38 (2010), p. 4050.

[2] http://www.guardian.co.uk/business/2007/jul/23/germany.greenbusiness; http://blog.cleveland.com/pdextra/2009/01/barack_obamas_speech_at_cardin.html

[3] *Renewable Energy Sources in Figures*, Federal Ministry for the Environment, Nature Conservation and Nuclear Safety (BMU), June 2010: http://www.erneuerbare-energien.de/inhalt/5996/42720/.

[4] *Renewable Energy Sources in Figures*, p. 10.

[5] *Renewable Energy Sources in Figures*, p. 18.

[6] *Renewable Energy Sources in Figures*, p. 27.

[7] See DECC, *Fast-track review of Feed-in Tariffs for small scale low-carbon electricity: Impact Assessment*, DECC 00059 (2010). http://www.decc.gov.uk/assets/decc/Consultations/fits-review/1439-fits-review-small-scale-cons-ia.pdf

[8]   *Renewable Energy Sources in Figures*, p. 29.

[9]   'Economic Impacts from the Promotion of Renewable Energies' (2010), 4049: Table 1.

[10] *Renewable Energy Sources in Figures*, p. 13.

[11] *Renewable Energy Sources in Figures*, p. 9.

[12] *Renewable Energy Sources in Figures*, p. 15.

[13] Monbiot, G., 'Solar PV has failed in Germany and it will fail in the UK', *Guardian*, posted 11 March 2010: http://www.guardian.co.uk/environment/georgemonbiot/2010/mar/11/solar-power-germany-feed-in-tariff

[14] References to Wolfgang Pfaffenberger's work can be found in 'Economic Impacts from the Promotion of Renewable Energies' (2010), but a convenient summary can be found in Pfaffenberger's contribution to the Renewable Energy Foundation's submission to the Stern Review: Pfaffenberger, W., 'Renewable Energy Policy In Germany: Experience and Problems. A Short Survey for The Renewable Energy Foundation', in Renewable Energy Foundation, *Submission to the Stern Review on the Economics of Climate Change*, 2006: http://www.ref.org.uk/attachments/article/167/ref.response.stern.review.0812.05.pdf.

[15] 'Economic Impacts from the Promotion of Renewable Energies', *Energy Policy*, 38 (2010), pp. 4048–4056.

[16] 'Economic Impacts from the Promotion of Renewable Energies' (2010), p. 4048.

[17] Manuel Frondel, Nolan Ritter, Colin Vance, *Economic impacts from the promotion of renewable energies: The German experience*, Rheinisch-Westfälisches Institut für Wirtschaftsforschung, October 2009.

[18] http://www.bmu.de/english/renewable_energy/doc/45291.php

[19] *Monitor*, programme number 613, 21 October 2010. See http://www.wdr.de/tv/monitor/sendungen/2010/1021/strom.php5

[20] 'Economic Impacts from the Promotion of Renewable Energies' (2010), p. 4050: Table 2.

[21] R&D figures from 'Economic Impacts from the Promotion of Renewable Energies' (2010), p. 4050.

[22] Frondel, M., Ritter, N., Schmidt, C. M., 'Germany's Solar Cell Promotion: Dark Clouds on the Horizon', *Energy Policy* 36 (11) (2008), pp. 4198–4202.

23 See, *Analysis of the Scope of Energy Subsidies and Suggestions for the G-20 Initiative*: *IEA, OPEC, OECD, World Bank Joint Report.* (16 June 2010.) Prepared for submission to the G-20 Summit Meeting Toronto (Canada), 26-27 June 2010. Available from: http://www.iea.org/weo/docs/G20_Subsidy_Joint_Report.pdf

24 *Analysis of the Scope of Energy Subsidies and Suggestions for the G-20 Initiative*, p. 4.

25 'Economic Impacts from the Promotion of Renewable Energies' (2010), pp. 4051: Figure 4.

26 'Economic Impacts from the Promotion of Renewable Energies' (2010), p. 4051. See Table 4.

27 ' Economic Impacts from the Promotion of Renewable Energies' (2010), p. 4053.

28 Traber, T., Kemfert, C. 2009. 'Impacts of the German support for renewable energy on electricity price, emissions, and firms', *The Energy Journal* 30 (3), pp. 155–178. Cited by 'Economic Impacts from the Promotion of Renewable Energies' (2010), p. 4053.

29 http://www.eprg.group.cam.ac.uk/wp-content/uploads/2010/08/Newbery1.pdf

30 'Economic Impacts from the Promotion of Renewable Energies' (2010), p. 4053.

31 Hillebrand, B., Butterman, H.-G., Bleuel, M., Behringer, J.-M., 'The expansion of renewable energies and employment effects in Germany', *Energy Policy* 34 (2006), pp. 3484–3494.

32 See 'Renewable Energy Policy In Germany'; see also Häder; M., Schulz, E., Beschäftigungswirkungen des EEG [Energie im Dialog, Band 6], Frankfurt a.M. 2005.

33 'The expansion of renewable energies and employment effects in Germany', p. 3493.

34 Rudyard Kipling, "A St. Helena Lullaby", in *Rewards and Fairies* (1910).

35 http://www.technologyreview.com/business/25016/

36 http://www.h-online.com/features/Solar-industry-creates-rocketing-demand-for-silicon-743401.html

37 http://www.solarserver.com/solarmagazin/solar-report_0806_chinese_e.html

38 http://www.e-to-china.com/2010/1226/89842.html

39 http://www.nytimes.com/2009/08/25/business/energy-environment/25solar.html?_r=1

[40] German Solar Industry Association (BSW-Solar), 'Statistic data on the German photovoltaic industry', June 2010, p. 2: http://www.solarwirtschaft.de/fileadmin/content_files/factsheet_pv_engl.pd f (2000–2009); Statistische Zahlen der deutschen Solarstrombranche (Photovoltaik), January 2011, p. 1: http://www.solarwirtschaft.de/fileadmin/ content_files/faktenblatt_pv_jan11.pdf (2010).

[41] http://www.polderpv.nl/PV_weltmeister_2010_prequel.htm#BNA1

[42] Peter Grösche and Carsten Schröder, 'Eliciting Public Support for Greening the Electricity Mix Using Random Parameter Techniques', *Ruhr Economic Papers*, No. 233, RWI (December 2010), p. 3.

Chapter Eight: Spain's Solar Eclipse

[1] San Miguel, G., del Río, P., and Hernández, F., 'An update of Spanish renewable energy policy and achievements in a low-carbon context' *Renewable Sustainable Energy* 2, 031007 (2010), p. 2; doi:10.1063/1.3301904. Available from http://jrse.aip.org/resource/1/jrsebh/v2/i3/p031007_s1.

[2] Drawn from data in 'An update of Spanish renewable energy policy and achievements in a low-carbon context', p. 2.

[3] 'An update of Spanish renewable energy policy and achievements in a low-carbon context', p. 4.

[4] Álvarez, G. C., et al., *Study of the Effects on Employment of Public Aid to Renewable Sources*, Madrid: Universidad Rey Juan Carlos, March 2009.

[5] President Obama, Speech at the Cardinal Fastener plant on 16 January 2009. http://blog.cleveland.com/pdextra/2009/01/barack_obamas_speech_at_ cardin.html.

[6] http://www.istas.net/web/abreenlace.asp?idenlace=6771

[7] Lantz, E. and Tegen, S., *NREL Response to the Report* Study of the Effects of Public Aid to Renewable Energy Sources *from King Juan Carlos University (Spain)*, Golden, CO: NREL, August 2009: NREL/TP-6A2-4621. http://www.nrel.gov/docs/fy09osti/46261.pdf.

[8] *Study of the Effects on Employment of Public Aid to Renewable Sources*, p. 27.

[9] 'The expansion of renewable energies and employment effects in Germany', p. 3491.

[10] 'Energías renovables situación y objetivos', Madrid: Ministerio de Industria, Turismo y Comercio, Gobierno de España, April 2010).

11 Blas, C., 'España admite que la economía verde que vendió a Obama es una ruina' [Spain admits that the green economy it sold to Obama is a wreck], *Expanción*, 21 May 2010, p. 34.

12 I am grateful to Annabel Ross who assisted me with this matter during her internship with Renewable Energy Foundation.

13 See Friends of the Earth, 'Recession leads to a fall in UK greenhouse gas emissions', 1 February 2011: http://www.foe.co.uk/resource/press_releases/ recession_ghg_01022011.html

14 DECC, Statistical Release, 'UK Climate Change Sustainable Development Indicator: 2009 Greenhouse Gas Emissions, Final Figures', 1 February 2011. http://www.decc.gov.uk/assets/decc/Statistics/climate_change/1214-stat-rel-uk-ghg-emissions-2009-final.pdf

15 'UK Climate Change Sustainable Development Indicator: 2009 Greenhouse Gas Emissions, Final Figures'.

16 http://www.iea.org/stats/pdf_graphs/ESELEC.pdf

17 http://www.xe.com/news/2010-11-23%2014:59:00.0/1541861.htm

18 http://www.xe.com/news/2010-11-23%2014:59:00.0/1541861.htm

19 http://www.bloomberg.com/news/2010-10-18/spanish-solar-projects-on-brink-of-bankruptcy-as-subsidy-policies-founder.html

20 http://www.renewableenergyworld.com/rea/news/article/2010/11/ spains-solar-power-sector-falls-into-the-abyss?cmpid=WNL-Wednesday-November17-2010

21 http://www.ecoseed.org/en/politics/laws-and-regulations/article/34-laws-regulations/8937-exclusive—spanish-solar-sector-to-intensify-subsidy-battle

22 http://www.ecoseed.org/en/politics/laws-and-regulations/article/34-laws-regulations/8937-exclusive—spanish-solar-sector-to-intensify-subsidy-battle

Conclusion: Green Prospects in the UK

1 http://www.gamesacorp.com/en/communication/news/gamesa-looks-to-establish-offshore-wind-technology-centre-in-scotland.html?idCategoria= 0&fechaDesde=&especifica=0&texto=&fechaHasta=

2 http://www.siemenshull.co.uk/

3 Webb, T., 'Siemens chooses Hull for wind turbine plant generating 700 jobs', *Guardian*, 20 January 2011. http://www.guardian.co.uk/business/2011/ jan/20/siemens-associated-british-ports-wind-turbines

[4] ROCs are traded bilaterally, so no detailed records are kept of all prices paid, but some traders publish useful data giving insight into approimate prices: http://www.e-roc.co.uk/trackrecord.htm

[5] See: http://www.hm-treasury.gov.uk/pespub_index.htm

[6] The REF database is provided free of charge, is fully searchable, and provides monthly output figures for all industrial scale generators registered under the Renewables Obligation: http://www.ref.org.uk/roc-generators/search.php.

[7] For example, the response of Mr Hendry, Minister of State for Energy, to a Parliamentary Question from Mr Philip Hollobone MP on 10 February 2011: http://www.publications.parliament.uk/pa/cm201011/cmhansrd/cm110210/debtext/110210-0001.htm#qn_o6. Mr Hendry reports departmental expectations of 14 GW of onshore wind and 13 GW of offshore wind by 2020.

[8] Department of Energy and Climate Change, *Estimated Impacts of Energy and Climate Change Policies on Energy Prices and Bills*, 2010, pp. 31–32. http://www.decc.gov.uk/assets/decc/what%20we%20do/uk%20energy %20supply/236-impacts-energy-climate-change-policies.pdf

[9] RenewableUK (formerly British Wind Energy Association), *Working for a Green Britain: Employment and Skills in the UK Wind & Marine Industries*, February 2011. http://www.bwea.com/pdf/publications/Working_for_Green_Britain.pdf

[10] *Working for a Green Britain*, p. 7.

[11] See press release for *Working for a Green Britain*: http://www.bwea.com/media/news/articles/pr20110201.html.

[12] http://www.bwea.com/media/news/articles/pr20110201.html

[13] http://www.statistics.gov.uk/cci/nugget.asp?id=285